Climate and Architecture

The Royal Danish Academy of Fine Arts
School of Architecture, Institute of Architectural Technology

Editor: Torben Dahl

Routledge
Taylor & Francis Group

Climate and Architecture

Editor: Torben Dahl
Coeditor: Winnie Friis Møller

Authors:
Nanna Albjerg
Torben Dahl
Eva Tind Kristensen
Nanet Mathiasen
Winnie Friis Møller
Georg Rotne
Peter Sørensen
Nina Voltelen
Ola Wedebrunn
Institute of Architectural Technology

The project was originally launched by
the late Professor Boje Lundgaard.

Graphics:
Rebecca Arthy and Cornelius Colding

Graphic design:
Jens V. Nielsen

Administration: Birthe Færch

First published (in Danish) by:
The Royal Danish Academy of Fine Arts,
School of Architecture Publishers, 2008

© The Royal Danish Academy of Fine Arts
School of Architecture, Institute of
Architectural Technology, 2008

This publication has been facilitated through support
from:
The Danish Ministry of Culture
Aase og Ejnar Danielsens Fond
Bergiafonden
VILLUM KANN RASMUSSEN FONDEN
Martha og Paul Kerrn-Jespersens Fond
Knud Højgaards Fond
The Royal Danish Academy of Fine Arts,
School of Architecture

First published 2010
by Routledge
2 Park Square, Milton Park, Abingdon, Oxon, OX14 4RN

Simultaneously published in the USA and Canada
by Routledge
270 Madison Avenue, New York, NY 10016

Routledge is an imprint of the Taylor & Francis Group, an
informa business

© 2010 Torben Dahl and The Royal Danish Academy of
Fine Arts, School of Architecture Publishers

Typeset in Univers by Alex Lazarou
Printed in Singapore by Markono Print Media Pte Ltd

British Library Cataloguing in Publication Data
A catalogue record for this book is available from the British
Library.

Library of Congress Cataloging-in-Publication Data
A catalog record has been requested for this book.

ISBN13: 978-0-415-56308-6 (hbk)
ISBN13: 978-0-415-56309-3 (pbk)

Contents

Preface

Many architecture books boast innumerable examples of new construction projects which in their different ways address the issue of energy saving and thus the highly topical issue of climate change. The subject of low-energy buildings, passive solar houses, solar energy utilisation and super-insulation is well documented in architectural journals and books. However, very few are based on an analysis of local climatic conditions or seek to explain the silent knowledge accumulated over generations in the way different architectural traditions are adapted to climate.

The neglect of climate as one of the key drivers in giving form to architecture has resulted in buildings today being characterised by conflict between globalisation's universal architectural expression and the contextual imperatives based on local cultural and climatic conditions. The adoption of the language of the ubiquitous steel or concrete framed glass box does not vary much whether the project is a hotel or office building in London or Kuala Lumpur. Whether located in tropical climates where shade and cooling are desired, or in arctic climates where heating and light are required, the solutions are generally dominated by the use of high energy-consuming climate control systems. These provide fresh air, appropriate levels of temperature and humidity, thereby ensuring comfort, but fail to exploit an alternative tradition based upon more responsive climatic design strategies.

Climate and Architecture addresses the architectural challenge of designing buildings, and especially their façades or climate screens, in order to maximise the potential of local climatic conditions and their associated construction traditions, in order to save energy and give users the means to control their own interior environment. Such an approach provides the means to elevate climate to its primary position as one of the major influences on architectural expression whilst also enriching the experience of occupying buildings.

This book explores the role climate plays in shaping the architecture of the world. Through an examination of examples from different climatic regions and different historical periods, the authors present a series of illustrated and expertly argued chapters. Various themes are explored from the physiology and psychology of light, heat, ventilation and comfort. The emphasis is upon the 'climate screen' – the interactive façade between inside and out – which adjusts to suit different weathers and seasons and personal tastes.

The book draws upon many contemporary and historic examples to present new perspectives on the power of climate to shape human habitation. The aim is to provide an overview of climate as one of the primary generators in giving form to architecture at its most fundamental level.

The book is written to appeal to students of architecture, practitioners and the general public. It deals with principles and their associated application in historic and contemporary architecture. The aim is to avoid presenting too much technical data (this is available elsewhere) so that the enduring interaction between climate and the design of buildings is fully explored.

Acknowledgements

The project group wishes to express its gratitude to all the authors and all the members of staff who have participated in the different stages of the project. A special word of thanks goes to our sponsors, Aase og Ejnar Danielsens Fond, Bergiafonden, VILLUM KANN RASMUSSEN FONDEN, Knud Højgaards Fond, Martha og Paul Kerrn-Jespersens Fond, the Danish Ministry of Culture and The Royal Danish Academy of Fine Arts, School of Architecture, for their support of the project.

Thanks are due to Winnie Friis Møller and Cornelius Colding for dedicated editorial assistance, and to Jens V. Nielsen for making his layout for the Danish edition available for the English edition.

Thanks are also due to Brian Edwards for his help in refining the text after translation and for adding valuable views, facts and information, and to Alex Hollingsworth at Routledge who fast-tracked the project to completion.

Torben Dahl
The Royal Danish Academy of Fine Arts,
School of Architecture, Institute of Architectural Technology

Fig. 0.1
LONDON CITY SKYLINE

The Architectural Potential of Climate

Torben Dahl

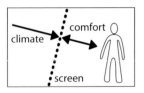

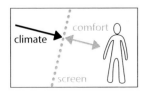

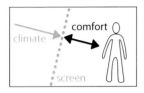

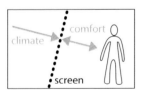

Fig. 1.1
CLIMATE AND COMFORT
This simple climate screen divides the world into an outside, where you are unprotected and exposed to climatic changes, and an inside, where you are protected and able to influence and adjust the circumstances that provide human comfort.

Human physiology enables us to determine the impact and assess the quality of our climatic surroundings. Our senses permit us – in an active interaction within individual parameters – to both experience and interpret impressions of climate and instantaneously control the state of our bodies in response to undesired influences. Buildings do not, however, work in this fashion.

The highly developed human adaptability mechanism, which in its original form has ensured man's survival, good health, comfort and pleasure, is not paid any particular attention by the regulations, design and construction practices that constitute the framework of modern architecture. As a result, the indoor climates they create exclude the outside world rather than respond and adapt to external conditions in a dynamic way.

Contemporary construction legislation governs a number of aspects, based on certain extreme conditions. This applies to the load-bearing structure, which is dimensioned according to maximum load. It also applies to the thermal insulation and dimensioning of the heating system, which is designed according to heat loss on the coldest night of the year, and it applies to engineering subjects such as hot water and ventilation. However, dimensioning of the indoor climate's ideal state is based on mean proportionals, average values and the absence of discomfort. As a result, our sensory apparatus is deprived of the stimulation that keeps it alive and active and enables it to register and react to extreme conditions in a positive manner.

The sophisticated climate control systems in modern buildings are designed to create and maintain an ideal level of comfort – one where as few individuals as possible feel discomfort because of ranges of temperature, humidity, draught, light and sound. Research into indoor climate has led to standardising the optimum level of indoor climate comfort and making it objectively measurable. An indoor climate, which – regardless of the nature of the outdoor climate – needs to correspond to 22 °C, having a 50% relative humidity and two air-changes per hour. This standardised indoor climate template implies that the appearance of buildings – interior as well as exterior – may also be standardised, thereby losing the potential for architectural expression found in the building's interaction with the local climatic context. The standardisation of regulations and comfort level has, until recently, led to a tendency to standardise architecture from the tropics to the arctic.

Research has begun to emerge that documents the benefits that individual influence on the indoor climate parameters, such as being able to turn the heating up and down, control the incident sunlight and open a window to let in fresh air, have on our sense of well-being. This suggests that responsiveness and personal control is more important than any objectively measurable levels of comfort.

It is therefore important to design the building envelope so as to enable this interaction with the user to occur. This dynamic between use, control and climate is common in vernacular architecture but has been neglected in much modern practice.

Fig. 1.2
CLIMATE SCREEN
The terraced houses in Skejby, Denmark, designed by Vandkunsten are a clear example of the climatic interplay between outside and inside.

Maintaining the standardised level of comfort – in terms of temperature, atmosphere and humidity – requires a lot of – often fossil-fuel based – energy. There may therefore be an untapped source of energy savings in letting the indoor climate follow the outdoor climate to a greater extent than what is determined by current standards and design methods.

The Earth's climate consists of a series of interactive systems, in which the individual climate parameters, such as heat, humidity, air movement and light, each contributes to the health of the whole with its own dynamic system. From each of these dynamic systems, principles may be derived which could serve as useful parameters in the design and environmental planning of a building.

If such parameters are used optimally, significant savings could be realised in the consumption of energy used to achieve a comfortable indoor climate, but additionally, the very notion of comfort may be explored and re-defined. This may contribute towards new approaches to architecture aimed at creating more pleasurable and stimulating buildings.

A detailed study of climate parameters, their interaction with the human body and with a building envelope may thus imply a potential for richer architectural experiences, increased well-being and lower heating bills.

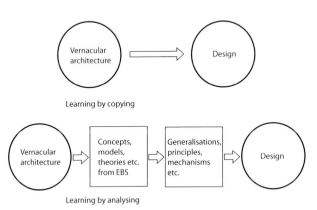

Fig. 1.3
EXPERIENCE FROM VERNACULAR ARCHITECTURE
In his article *Vernacular design as a model system,* Amos Rapoport indicates the appropriateness of carefully studying underlying experience and theories through so-called EBS (Environment-Behaviour Studies), thereby deducting principles from vernacular architecture so that it can be used as a model.

A study of traditional architectural customs offers a wealth of climatic understanding and an inexhaustible source of inspiration for climatically adapted architecture. A traditional architectural custom anywhere on the globe is characterised by its ability to ensure an optimised level of resource saving and climatic adaptation and, consequently, a high level of built-in sustainability.

It was a point noted 2,000 years ago by the Roman architect Vitruvius in the 6th book of his *The Ten Books on Architecture*:

If our designs for private houses are to be correct, we must at the outset note of the countries and climates in which they are built. One style of house seems appropriate to build in Egypt, another in Spain, a different kind in Pontus, one still different in Rome, and so on with lands and countries of other characteristics.

Throughout the world, the expressions of traditional architecture are based on and adapted to local conditions. This applies primarily to the local availability of materials and the response to climatic conditions. However, the distinctive cultural, religious and social character of a neighbourhood or region may also influence the design.

Today, climatic adaptation is practically synonymous with energy reduction and CO_2 savings. For this reason, the contemporary architectural discourse is focused more on sustainability in terms of minimising energy consumption than on ensuring an optimum utilisation of and adaptation to the outdoor climate. This seems too narrow a response to energy related design and ignores the potential of greater attention to climate architecture.

As examples of climatic control with a particular focus on a reduction of energy consumption, it is worth mentioning the German-Austrian Passive House Concept, the British high-tech glass and steel constructions as well as more humble and more widely resource-conscious low-tech initiatives. The three examples shown here represent these categories.

Each of the examples contains significant elements in which adaptation to the local climate forms a part to varying degrees. In the first two examples, supply and control technology dominate, whereas spatial design and utility differentiation characterise the last. Nevertheless, it seems that none of the three examples adequately incorporate climatic experiences from local, traditional architectural customs.

The Passive House concept, Vorarlberg

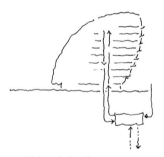

High-tech, London

Low-tech, London

Fig. 1.4
ENERGY-SAVING CONCEPTS

Fig. 1.5
GEMEINDEZENTRUM LUDESCH, VORARLBERG, AUSTRIA, 2005
Architect: Hermann Kaufmann

The project is a model project for the Passive House concept, featuring low levels of very well-documented energy consumption and a controlled used of materials. The exterior walls are highly insulated by means of 35 mm insulation material, and the technology of the heating and ventilation system is complex, consisting of heat pumps from groundwater used for both heating and cooling, supplemented by hot water from both solar panels and a biomass heating system and equipped with a heat recovery and extraction system for the ventilation air. The project has a consistent use of local wood and a traditional high quality of craftsmanship. The Passive House concept involves the idea of minimising the interaction with the external climate to save energy, which on one hand makes the idea globally applicable, but on the other hand it shows no particular consideration for local architectural culture regarding climatic adaptation and utilisation.

Fig. 1.6
CITY HALL, GREATER LONDON CITY COUNCIL, LONDON, 2002
Architect: Foster and Partners

An important starting point for the design of this building situated on the southern bank of the River Thames was the building's uncompromising design in relation to environmental conditions – wind, daylight levels, sun path and heat flows. The energy strategy planned for the City Hall has resulted in an energy consumption that is only one quarter of the energy used by a typical office building. This has been achieved by means of both passive environmental systems and a careful design giving consideration to energy loss and reduction of incident light. As a result, London City Hall stands as a significant example of the integration of advanced technology combined with the utilisation of local geography and natural climate control potential.

Fig. 1.7
STRAW HOUSE, STOCK ORCHARD STREET, LONDON, 2004
Architect: Sarah Wigglesworth & Jeremy Till

The principal idea of the house is to create a model for a sustainable life in the city based on low-tech materials and systems. The Straw House is made of materials chosen primarily on the basis of intuition without great charts and long calculations. The building design has a strong focus on minimising energy in the utility phase. Passive solar energy is gained from the large south-facing window apertures, and in the walls to the east, north and west, extra insulation has been used, in which the U-values are considerably below those stated at the time in UK construction regulations. The building's main floor is divided into three sections. An office space in two adjoining floors; sleeping quarters insulated by means of walls made of bales of straw, and in between these two sections, an open space with large window apertures to the south. The space is used for short periods of time, and the changing temperature that results from the relatively poor insulation can be tolerated and even enjoyed as a positive experience.

CLIMATE AND COMFORT

Place and Climate

Human Comfort

Fig. 2.1

Place and Climate

Peter Sørensen
and Winnie Friis Møller

All architecture is affected by climate, partly because a building needs to protect its interior against exterior climatic influences, partly because the building needs to be protected against erosion caused by climate. The interaction between place and climate is of critical importance in architecture and of equal importance to the sense of place. Architecture is a connecting link between place, climate and human life.

Attention to climate can be studied in any building in major or minor detail, but climate-adapted architecture is primarily found in traditional architecture, in which it is evident that the tougher the climate, the more characteristic and distinctive the resulting architectural forms. Architecture is as a whole and in detail designed through the experience gained from a long development process based on the resources of the specific place and its particular climatic and cultural conditions. By studying traditional architecture and the principles of climate adaptation and control found in vernacular buildings, it is possible to understand and exploit the hidden knowledge and experience behind the design as an inspiration for contemporary architecture. Traditional architecture can be an important design guide, especially in the area of offering clues to how to respond to particular local climate conditions.

WEATHER AND CLIMATE

Whereas the concept of weather covers the immediate atmospheric situation in a given place on Earth, the concept of climate indicates the average weather over a number of years, usually at least 30 years. Thus, climate changes are registered slowly.

The word climate has its origins in ancient Greek and denotes the inclination of Earth in relation to the rays of the Sun. As the word implies, understanding of the climate was based on a direct relationship between the incoming solar radiation and temperature. The Greek astronomer, geographer and mathematician Ptolemy (c. AD 90-168) divided Earth into climate zones on the basis of air temperature, which he believed followed the angle of the incident sunlight. Consequently, he assumed that the highest temperatures on Earth would be found around the Equator, and that temperatures would then gradually decrease towards the poles. This scale was divided into three climate zones; a warm zone around the Equator where the angle of incident sunlight is at its maximum; a temperate intermediate zone; and a cold zone by the poles where the angle of incident sunlight is at its minimum.

Vitruvius bases his ideas on the same three climate zones on Earth, created by the Sun's path across the sky:

… one part of the earth is directly under the suns's course, another is far away from it, while another lies midway between these two … we must make modifications to correspond to the position of the heaven and its effects on climate.

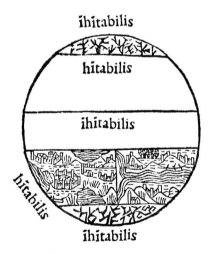

Fig. 2.2
HABITABLE CLIMATE ZONES
In the 13th century, the British scientist Sacro Busto described five climatic zones on Earth on the basis of Ptolemy's spherical geometry. He only considered two of the zones fit for human habitation.

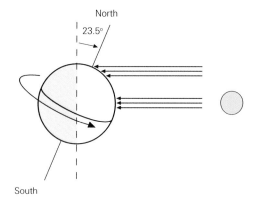

Fig. 2.3
ANGLE OF INCOMING SOLAR RADIATION

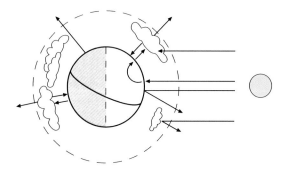

Fig. 2.4
EARTH'S RADIATION BALANCE
The amount of clouds in the atmosphere is of great significance to the Earth's radiation balance. If there are a lot of clouds, a considerable amount of the incoming solar radiation will be reflected or absorbed and therefore not reach the Earth's surface. At the same time, a large proportion of the outgoing radiation from Earth will be sent back towards the Earth's surface as atmospheric counter radiation, contributing to the heating of the atmosphere (the greenhouse effect).

Today, weather and climate are defined by a whole string of measurable parameters, the primary ones being:
– Solar radiation – significant to both temperature and lighting conditions
– Wind and air pressure conditions
– Humidity and precipitation.

The foundational principles in these primary climatic factors and systems form the basis for climate control, whether in relation to the Earth's climate as a whole, climatic conditions in and around the individual building, or man's concept of comfort.

GLOBAL DYNAMIC WEATHER SYSTEMS
The Earth receives practically all its energy from the electromagnetic radiation of the Sun. This energy drives a series of weather phenomena in the Earth's atmosphere, which combines to form a dynamic system in which the individual parameters are closely interdependent and in constant change. An understanding of these systems and their dynamic interplay is important in order to understand the weather and climate correlations that affect both buildings and people. These systems also constitute a framework for the climatic characteristics and conditions that are prerequisites for climate adaptation and climate control of buildings in any given place.

Earth's radiation balance
Air temperatures at the Earth's surface are, above all, determined by that place's radiation balance. The concept describes the relationship between the amount of shortwave incoming radiation from the Sun that hits the Earth's surface and the amount of long wave radiation that the heated surface of the Earth emits back to the atmosphere. The greater the incoming radiation, the more the Earth's surface is heated, the greater the outgoing radiation, the more it is cooled.

However, the amount of incoming solar radiation that reaches the Earth's surface varies from place to place. Three primary factors in the relationship between the Sun and the Earth influence this: the angle of incoming radiation, the time of incoming radiation and the atmospheric reduction of the radiation.

The angle of incoming radiation is determined by the place's location on Earth in relation to the altitude of the Sun. The amount of solar radiation that hits the surface of the Earth per m² is greatest when the Sun is at its zenith. The time of incoming radiation depends on the length of the day. Atmospheric reduction is due to particles or water vapour in the atmosphere that either absorb the radiation or reflect it back into space. The majority of this reduction is caused by clouds although there are also human factors involved.

Consequently, the cloud cover in a given place is decisive for the place's radiation balance and thus for temperature, climate and design. Because of the shorter distance through the atmosphere, the atmospheric reduction decreases in relation to the altitude above sea level. High-lying areas will often have a higher level of incoming radiation than low-lying areas, but as the air is also cooled in relation to the altitude, air temperatures in mountain areas will drop by approx. 10 °C for every 1,000 m altitude increase.

EARTH'S PRIMARY WIND AND PRESSURE SYSTEM

When air mass is heated, it expands, which causes it to rise. This concept is called thermal lift. When air mass rises, a low pressure is created at the surface, whereas dropping air mass will create high pressure at the Earth's surface, as the air here is compressed. The very uneven heating of the Earth's surface caused by the great differences in the amount of incoming solar radiation is intensified by the different abilities of land areas and oceans to absorb heat. The air temperature at the Earth's surface therefore varies greatly between different zones on Earth, which leads to pressure differences. This creates a system of convection currents in the atmosphere, which constantly attempt to even out temperature and pressure differences. As the Earth rotates, winds are deflected, and bands of primary wind directions are created. This results in different climates in different places at the same latitude and in different climates at different seasons. Buildings, being largely fixed objects, have the task of adapting to these variations.

The primary wind and pressure system moves in relation to the Equator through the year, as it is symmetrical around *the intertropical convergence zone* (ITC), which follows the Earth's maximum solar heating zone with approx. one month's delay. As a consequence, most areas on Earth experience seasonal variations in wind directions and precipitation. Special climate conditions even derive their names from these primary wind conditions, e.g. the monsoon. Special building traditions also emerge as direct responses.

Earth's cloud cover and precipitation

Clouds are primarily formed by air heated at the Earth's surface rising and condensing into water vapour in the form of cloud formation by cooling. When the amount of water vapour reaches its saturation point, precipitation occurs. Because of the strong heating caused by the direct incoming solar radiation at ITC thermal low pressures are created, which cause a band of clouds around the Earth with many small clouds. The rising air mass releases precipitation, and land areas are primarily covered by rain forest whose dense plant life retains the humidity. Study of traditional architecture in such warm, humid climate zones as the Amazon basin shows how modern design could achieve greater climatic responsiveness without being totally dependent of air-conditioning.

To the north and south of these areas lies a belt with a series of dynamic high pressures. Here, there are relatively few clouds to reduce incoming radiation, and as a result, day temperatures are very high.

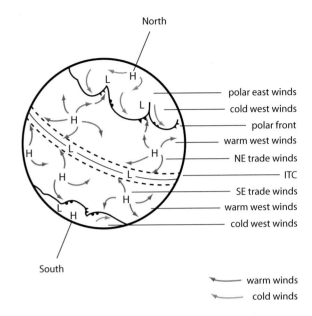

Fig. 2.5
EARTH'S PRIMARY WIND AND PRESSURE SYSTEM

In these areas, the warmest and driest areas on Earth are found. The missing cloud cover results in significant heat radiation into space at night, which causes great temperature variations between day and night. So although we think of forming buildings solving air-cooling as a major problem, at night the drop in temperature results in necessary attention to insulation.

Around the polar fronts, the cloud cover increases again, as in this area, there is a band of wandering low pressures around the Earth, which occur when hot and cold air meet. Characteristic spiral cloud bands are often formed around the low pressures. Coastal Denmark is highly influenced by these weather conditions, which cause a lot of precipitation and unstable weather. This also results in local differences such as architecture of Western Jutland being different in character to that of Eastern Zealand.

The North and South poles are marked by constant thermal high pressures because of the cold, and as a consequence, there is relatively little precipitation in these areas. The same applies during winter to the interior of the large continents in the northern hemisphere, where the cold causes thermal high pressures.

CLIMATE CHARACTERISTICS

On the basis of the global weather systems, it is possible to define a number of characteristic climate situations that are of great principle significance to regional climate-adapted architecture. In the interplay between climate and building design it is important to know both normal and potential, more extreme climate conditions. Are the variations in temperature, precipitation and wind only minor or are there great variations seasonally or between day and night? Are there any predominant wind directions or other specific seasonal climatic phenomena to take into consideration?

Most commonly, traditional buildings are designed according to the climatically worst time of the year, whether this be summer heat or winter cold, in terms of thermal climate. However, it may also be the rainy season that determines the design of a house – maybe at the expense of general usability throughout the rest of the year. Certain areas are marked by predominant wind directions or characteristic seasonal winds, such as the monsoon, the sirocco etc., which may also have left their mark on local architecture.

In the following sections, a number of characteristic climate types are described with examples of how traditional architecture is inspired by and designed according to the climate. The grouping has been made primarily according to temperature into a *hot*, *moderate* and *cold* climate, respectively, and secondarily according to humidity into a *humid* or *dry* climate, respectively.

Hot – hot humid/hot dry

Moderate – moderately dry/moderately humid

Cold – cold humid/cold dry.

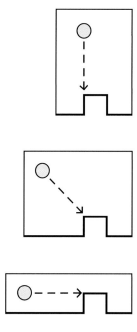

Fig. 2.6
ALTITUDES OF THE SUN
The three vignettes characterise the climate zones hot, moderate and cold.

Hot humid

Tropical

Stable high humidity and temperature, no defined seasons and only little climatic difference between day and night. In the rain forest, humidity is constantly high with precipitation throughout the year. The heavy cloud cover reduces radiation, but may also cause a strong glare.

Buildings in these areas are typically open skeleton structures with large roofs to protect against and divert rain. Most often, building materials are lightweight materials such as bamboo, fibres or leaves. The light walls ensure that the houses are well ventilated and make the best possible use of wind for ventilation. Floors and shaded terraces are raised above the terrain to protect against flooding, moisture and small animals and to provide ventilation through the floor. Houses may also be placed on water to make maximum use of the cooling caused by evaporation. Outdoor kitchens remove heat, air, insects, rodents and fire hazards from the buildings.

Example of a building type adapted to tropical humid climate: house on stilts, Thailand.

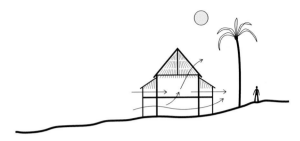

Fig. 2.7
THAILAND – HOUSE ON STILTS

Subtropical

Humid, warm coastal climate outside the tropics with long, warm and humid summers and short winters, often with strong, cool winds.

Buildings are primarily constructed to protect against the humidity and heat of summer, and secondarily to protect against cool winds during the short winter. Often they feature ventilated, high-ceilinged rooms, spanning from one façade to another. Large, covered terraces protect against sunshine and rain and provide opportunities for staying outdoors. Apertures protected by louvres allow air in and keep rain out. Interiors are light and open in order to avoid gathering moisture and heat. Buildings are made of light materials such as wood and metal and often surrounded by gardens with water and shading trees. Foundation stones raising the upright wooden poles above the moisture of the Earth are a typical architectural detail from the humid climate.

Example of a building type adapted to subtropical humid climate: a traditional Japanese house.

Fig. 2.8
JAPAN – TRADITIONAL HOUSE

Hot dry

Desert

Extreme heat, fierce sunlight, clear sky, rare or no precipitation and a lack of water. Mild winter and extremely hot summer. Major difference between day and night temperatures.

The nomads of the desert area deal with extreme climate conditions by using lightweight transportable materials. The Bedouin tent provides maximum shade and ventilation, protects against night cooling towards the cloudless sky, and

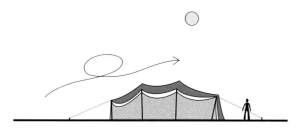

Fig. 2.9
SAUDI ARABIA – BEDOUIN TENT

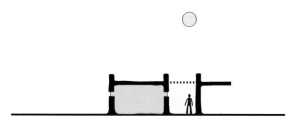

Fig. 2.10
MOROCCO – MUD HOUSE

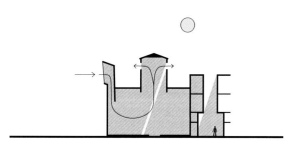

Fig. 2.11
EGYPT – TOWN HOUSE

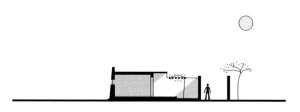

Fig. 2.12
SANTORINI – TOWN HOUSE

creates shelter and protection against sandstorms and the desiccating effect of the wind. Black woven wool blankets filter light and air, protect against the glare of the sunlight and become dense and cooling when moistened.

Permanent residents in desert-like areas use great mass and heavy materials to delay the effect of the huge temperature variations of the day. Subterranean dwellings provide maximum protection against extreme temperatures, and buildings are often designed as climatic imitations of these with thick mud walls or unbaked brick, small protected apertures and small shaded courtyard spaces.

Roof terraces are used as living areas during the night. In areas where cool winds are constant, as e.g. in Iran and Egypt, they are led into the building. Water cooling is used in numerous ways both outside and inside. Enclosed courtyards achieve a comfortable microclimate through the use of water and plants. Lightweight materials are used to provide shading cover.

Examples of a building type adapted to extremely hot and dry desert climate: Bedouin tent, mud houses in Morocco, Egyptian town house.

Moderately dry
Mediterranean climate
Long, warm summers and short, cool and humid winters. The climate is mild with little variation between day and night.

Traditional buildings are heavy stone buildings, which provide a stable indoor climate throughout the long summer, but require tolerance and heating during the short winter. Buildings are often whitewashed on both the inside and the outside. In order to keep the strong sunlight out, there are only a few, small windows. Shutters and louvres provide extra shielding against heat and light during the after-noon siesta. Loggias, balconies, terraces, porticos, patios, enclosed courtyards and gardens provide buffer zones for outdoor living during the summer.

Examples of building types adapted to a Mediterranean climate: town house from Santorini, Roman atrium house.

Continental climate
Long, warm and dry summers and long, cold winters. Significant seasonal temperature differences and temperature variations during the day. Low humidity and strong winds.

The North American plains Indians' conical tepee is a good example of a climate-inspired nomad dwelling. The tent is erected with the entrance in the steep, strengthened back against the wind. Adjustable smoke flaps are used to control ventilation through the tent top. Because of its aerodynamic design, the tepee is surprisingly comfortable, even during winter, due to the use of a central fireplace and an extra animal skin lining.

The traditional Turkish dwelling handles the need for climate adaptation for both the warm summer and the cold winter by using heavy materials for the substructure and a superstructure made of an insulating timber structure, stone faced roofs with a low pitch and overhangs that protect against the summer sun. Transitions and living areas are differentiated between inside and outside, and central, heated rooms on the top floor provide protection against the cold and windy winter.

Example of building type adapted to a continental climate: Turkish country house.

Moderately humid

Temperate coastal climate

Moderately warm, humid, rainy and changeable. This climate type covers coastal areas approximately halfway between the Equator and the North and South poles. The climate is characterised by its proximity to the sea, often windy, with frequent precipitation and short, cool summers and mild winters.

Buildings are protected against humidity and wind. They may be orientated so that the gable faces towards the sea and the prevailing wind, and they may have porches and unheated rooms as a transition to heated rooms, which are normally gathered around a chimney – the heart of the house. Frequent precipitation has led to a high pitch and roof overhangs that are either clad in pipes or straw to divert water and insulate, as is seen on the Danish island of Fanø, or clad in more lasting stone materials, as in Ireland. Walls are most often made of bricks. Windows can be small to protect against cold and winds in particularly exposed areas or large to compensate for the sparse daylight during dark winter periods as in Scandinavia, Great Britain and the Netherlands. Buildings along the coast of Southern England, Northern France and Northern Spain are often protected against the wind by means of a light structure with a lot of glass, which forms a climatic buffer zone.

Examples of building types adapted to a temperate coastal climate: rectangular-shaped house, Fanø, Denmark, and town house in La Coruña, Spain.

Cold humid

Subarctic

Cool summers, bright nights with short, intense growth periods, cold, dark winters, long snow-covered periods. Changing between quiet, dry periods and humid periods and very windy weather.

The large, continuous Scandinavian, Siberian and Canadian coniferous forest areas have a continental climate with long cold and windy periods. Here, the fireplace is a central element. The nomadic Saamis erect their tents in protective areas near forest and water with the door opening facing east. Two parallel planks with the fireplace in the centre divide the room into a living and a sleeping area with thick layers of birch rod and reindeer skins as insulation against the cold. The tent, which has a smoke opening at the top like the American tepee, encircles the fireplace's flames and radiant heat, whilst sides and back are protected by means of insulating animal skins.

Fig. 2.13
ROME – ATRIUM HOUSE

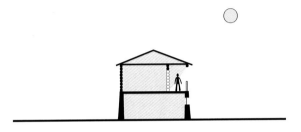

Fig. 2.14
TURKEY – COUNTRY HOUSE

Fig. 2.15
FANØ – RECTANGULAR-SHAPED HOUSE

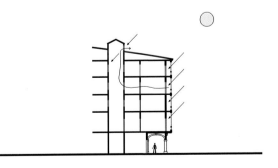

Fig. 2.16
LA CORUÑA – TOWN HOUSE

Fig. 2.17
LAPLAND – SAAMI TENT

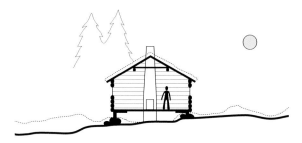

Fig. 2.18
SWEDEN – LOG HOUSE

Fig. 2.19
ICELAND – PEAT HOUSE

The Swedish forest protects against the wind during winter. When snow settles on a roof with low slope it helps to retain the heat in the house. Here, there is sufficient timber for building massive walls around the heavy, brick chimney core at the centre of the house, which retains heat for a long time. By way of contrast the Icelandic peat house finds no shelter in the bare land but is built down into the terrain, exploiting the Earth's moisture, air and thermal absorbing warming qualities by means of thick peat walls and grass roofs. Internal wood panels on floors, ceilings and walls reduce heat radiation, thereby increasing the room's surface temperature and breaking thermal bridges which tend to occur. The individual rooms have a permanent layout according to their specific function, and furniture is often an integrated part of the room's permanent internal fittings.

Examples of building types adapted to the humid, subarctic climate: Saami tent, Swedish log house, Icelandic peat house.

Cold and dry

Arctic

Extreme cold, long, dark winters and short, bright, cool summers. The temperature changes very little between days of 24 hours' darkness and days of 24 hours' light.

As is the case in the extreme heat of the desert, in the extremely cold areas, the severity of the temperature is critical for human survival. Protection against cold and particularly the cold wind is the greatest challenge. Animal life has inspired man's survival strategies in these areas, and animal skins and furs are used for clothing and insulation of dwellings.

The snow hut, i.e. the igloo, demonstrates man's ability to develop architecture that provides maximum protection in extreme climatic conditions. The semi-circular structures have a minimum surface in relation to volume; they are constructed by means of snow blocks, which are full of air and therefore insulating. With a sunken entrance tunnel and internal lining, body heat and a small heat contribution are sufficient to achieve indoor temperatures above 15 °C, even at very low outdoor temperatures.

The yurt, an advanced, portable dwelling for nomads in Siberia and Mongolia has a similarly optimum shape, surrounding a central fireplace.

Examples of building types adapted to extremely cold climate: Eskimo igloo, yurt – Mongolian nomad dwelling.

PRINCIPLES FOR CLIMATE ADAPTATION

A building's capacity to adapt to or utilise climate is a consequence of the way in which the building is constructed in relation to climate and surroundings, materials and resource consumption. The adaptation ability depends on both permanent, passively climate-controlling parts and changeable, actively climate-controlling parts. It may be the building's heavy or light structures and its spatial layout, or it may be transitions between inside and outside in terms of façade or window design that contribute actively to responding to and adjusting climate between inside and outside.

In principle, there are three main ways in which a building may be climate-adapted in relation to its use:

a. A passively climate-controlling building is unchangeable, but the building and its spaces can be used in various ways over day and night or at the different seasons of the year in relation to the changing climate.
b. An actively climate-controlling building can change dynamically and adapt to changing climate conditions.
c. A building that combines the principles of both a and b, can both be used in varied ways and actively adapt to the climate.

In the passively climate-controlled building (a), rooms are formed to be generally usable but have different climatic adaptation in terms of layout, choice of materials and construction, and can be used differently in relation to the hours of the day, the seasons and the changing climate.

In the actively climate-controlled building (b), rooms can adapt to different climatic situations through an active adjustment of light, air and temperature by means of an interactive façade, e.g. a buffer zone, adjustable shutters and blinds.

A combination of passive and active means through climatically varied rooms (thermal mass), temperature zoning (insulation) and actively climate-controlling façades (utilisation of passive solar energy) can provide an optimal solution (c).

Using new technologies, methods of construction and materials, it is now possible to make our buildings much more flexible and climate active, so that they can adapt to changing climates to a much higher degree through the day and through the seasons. If these new possibilities are combined with experience from the traditional examples, this may result in both a more beautiful and a more functional, interactive and energy efficient architecture. A new design strategy like this is also needed to provide low energy buildings more able to adapt to changing conditions under the impact of climate change.

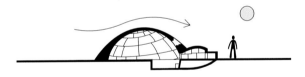

Fig. 2.20
THE ARCTIC – ESKIMO IGLOO

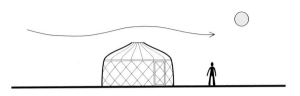

Fig. 2.21
MONGOLIA – YURT

Human Comfort

Torben Dahl
and Peter Sørensen

Shelter and protection against the weather is the earliest and primary function of the house. As a warm-blooded being, man needs shielding from the extremes of weather in order to survive and feel good or comfortable.

The concept of comfort has its linguistic origin in French (*confort*), but is fully integrated into all European languages both in everyday use and as a scientific term. But the term has developed over time. For example, in the Danish *Salmonsen's Encyclopaedia* from 1923, the comfort concept is defined as 'homely cosiness, caused by the home's practical and stylish interior design'. However, the English definition in Webster's *Concise Dictionary*, 1998, is much closer to contemporary use. Here, the concept of comfort is defined as follows: *A state of ease and satisfaction of bodily wants*. A state, where the body not only finds protection against the elements, but also experiences well-being and satisfaction.

'Indoor climate' is the technical term for the climate created by the house's physical enclosure and the various climate systems employed. However, although comfort is the objective, different quality levels can be established for indoor climate.

An acceptable indoor climate is one which ensures the absence of damaging effects on health and ensures that at least 80% of residents or users of buildings are satisfied. The 'acceptable indoor climate' is typically the basis for legislation for both housing and in relation to working environment. A 'good indoor climate' takes individual needs into account, facilitates personal influence on environmental conditions and considers particularly the needs of sensitive and exposed groups. An 'excellent indoor climate' offers positive stimulating effects, well-being, experience and variation above the two earlier climatic standards.

The psychologist Abraham Maslow is known for his description of human needs in the so-called 'Maslow's hierarchy of needs' (see Fig. 3.2). His claim is that needs must be met from the bottom of the pyramid and upwards. So, the need for food, water, comfort etc. must be met, before you can start meeting the need for physical and social security. According to Maslow, a need is a congenital or later acquired desire aimed at satisfying physical, psychological or social demands. The five steps of Maslow's pyramid from the bottom upwards are:

Physiological needs – such as consumption of food, water, air, physical movement and reproduction.

Safety needs – security, stability, order, protection against the elements, pains and other discomforts, and the absence of anxiety and fear.

Social needs – need for belonging, love and friendship.

Esteem needs – at this level, the needs concern self-respect, confidence, achievement, recognition, status and dignity.

The need for self-actualisation – at the top level, you seek to actualise congenital or later acquired abilities, to reach so-called peak experiences or revelations. (source: Wikipedia).

Fig. 3.1
VALS
Baths in Vals, Switzerland, 1996.
Architect: Peter Zumthor.

Fig. 3.2
MASLOW'S HIERARCHY OF NEEDS

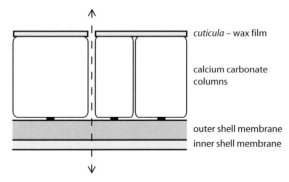

cuticula – wax film

calcium carbonate columns

outer shell membrane
inner shell membrane

Fig. 3.3 a-b
EGGSHELL LAYERS
The eggshell consists of a series of multifunctional layers that combine to form an effective climate-controlling membrane.

Maslow considered the three lower levels to be deprivation needs. He considered the two upper levels to be growth needs. The deprivation needs can be satiated, and in that sense they are in contrast to the two upper needs, which cannot be satiated. Normally, the deprivation needs are the strongest, which means that the desire for esteem and self-actualisation recede into the background if one or more of the deprivation needs are not met.

MEMBRANES

Consider an egg as an example of a closely cocooned and safe haven – an enclosing shell around a soft centre, a concept of well-being, protection and ease, where the basic needs of a developing being are met.

The egg concept can be further analysed as an example of the fulfilment of the physiological and safety needs in relation to building construction and climate. If you take a closer look at the fine structure of an egg and compare the eggshell to the envelope of the house, you find answers to almost all of the functional requirements that need to be met in building by the climatic enclosure.

A section through an eggshell shows first a strongly alkaline wax film (*cuticula*), which makes the new-laid egg shiny, and which is known to have an antibiotic effect. Then follows a layer of calcium carbonate 'columns', which form the hard part of the eggshell. This is perforated by fine pores, which allow the exchange of oxygen, carbon dioxide and moisture, but not bacteria. The layer of lime contains pigments that – regardless of whether the egg is dark or light – reflect the heat of the Sun and are able to keep the egg's temperature below 30°C even during extended periods of solar influence. This is important to the survival of the foetus. On the inside of the shell, there are two membranes of slightly different thicknesses, which both serve as physical and chemical barriers. They also contain anti-bacterial substances. In the course of the 21 days that pass before the chicken breaks out, the eggshell of a hen's egg will be penetrated by 6 litres of oxygen, 4.5 litres of carbon dioxide and 11 litres of water vapour.

The eggshell's climate-controlling and disinfecting effect is utilised by the bushmen of South Africa, when they store drinking water in emptied ostrich eggs buried in the ground, where they can stay for a long period of time as water depots without the water being spoilt.

Human skin

Like the eggshell, human skin is a multi-functional climate screen with a number of passive and active functions, which could be termed our personal protective packaging against the surrounding world.

The skin's passive functions include:
– Protection against cold, heat and radiation.
– Protection against pressure, bumps and ripping.

– Protection against the effect of chemical substances.
– Protection against penetration by germs, mainly through the development of an acid protection layer.
– Protection against heat and water loss.

The skin's active functions include:
– Preventing penetration by micro-organisms.
– Absorption of certain biologically active materials.
– Secretion of sweat, cooling function; along with sebaceous glands, production of hydro-lipid film.
– Circulation and thermoregulation through blood supply to the skin.
– Sensing of pressure, vibration, touch, pain and temperature.

In the main, human skin represents all of the functions that need to be performed by the building's envelope or what we here call the buildings climate screen, but the skin – like the eggshell – has a very dynamic interchange with the surroundings with regard particularly to heat, moisture and biological material. The skin is also very sensitive towards broader changes in the surrounding climate.

As a contrast, the technological development of the typical building façade has been aimed at preventing all interchange, making the façade elements completely tight, thus achieving maximum control of the indoor climate.

However, under the impact of sustainable design the trend is now more towards a dynamic approach to the construction of climate screens, aimed at ensuring optimum interplay between inside and outside.

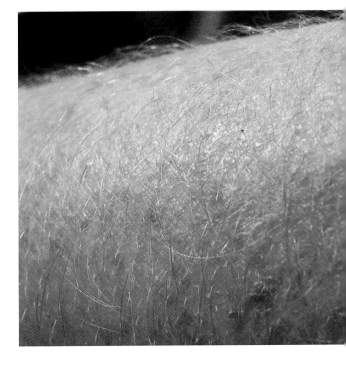

Fig. 3.4
THE SKIN AS CLIMATE SCREEN

The intelligent climate façade – SmartWrap™

In their SmartWrap concept, the American architectural firm KieranTimberlake Associates have suggested a way in which to integrate the individual functions of a conventional external wall into an advanced composite material, which can envelop the architectural structure and hence contain the building's function more effectively.

The concept is demonstrated in an exhibition pavilion in order to show the architectural possibilities and the associated construction physics technologies.

The basic material is a polyester film that protects against rain and wind whilst at the same time serving as a substrate for the other SmartWrap™ layers. Polyester is known from the packaging and clothing industries and has been chosen because it is inexpensive, colourless, transparent, quick-drying and mildew-resistant. Additionally, it does not absorb moisture, it has great mechanical strength, and function layers can be applied by means of e.g. inkjet printing and roll-coating.

In order to control the thermal climate, a phase change material is applied to the substrate, which absorbs excess heat and releases it again when temperatures drop. Furthermore, the substrate is equipped with a layer containing organic

Fig. 3.5
SMARTWRAP™
The exhibition pavilion designed for an
exhibition at Cooper-Hewitt, National
Design Museum in New York in 2003.
Architect: KieranTimberlake Associates.

light-emitting diodes driven by a layer of organic photo-voltaics, which can also
generate electricity for the building.

Because of this many benefits the architectural firm calls SmartWrap™ the
climate screen of the future. Its design and production mimics many biological
functions and has the benefit of being flexible and mass customisable.

MAN'S ADAPTATION TO CLIMATE

The body and the skin's sophisticated protection and adjustment functions combined with clothing make it possible to survive under many different and even
extreme climatic conditions. The human body has a very large capacity for thermal
well-being – maybe from minus 20 °C to around plus 40 °C, as the experience of
heat/cold is highly affected by the other climate parameters – humidity, wind velo-
city and heat radiation – and the body's movement and clothing are also of decisive
significance. Everybody recognises the sauna situation of dry air around 80°C,
which is experienced as positive – albeit for a limited amount of time – and the
semi-undressed sunbathing in front of the mountain hotel in quiet, dry air and
temperatures below freezing.

However, although man is capable of surviving in almost any climatic situation
around the world, as long as he can find food, the climatic conditions in which man
can develop physically and mentally are more limited. The normal situation may be
described as a biological battle for a state of equilibrium in relation to temperature,
humidity and air movement, in which man uses a minimum amount of energy in
order to adapt to his surroundings.

Comfort zones

Research in this field has identified the climatic conditions under which man is
most comfortable, both at work and while resting. A temperature range of 22 °C
± 2 °C is considered an operational comfort level in the ambient air.

The human body has a relatively high degree of tolerance to the air's relative
humidity, and it can stay in 20–80% relative humidity without any noticeable discomfort. The body is also tolerant to wind and movements in the air.

However, the complex interplay of these three important parameters – temperature, humidity and air movement – in both outdoor and indoor climate combined
with clothing and nutrition makes it a scientific question to identify optimum conditions for extended periods of stay and work.

The standardised comfort concept

The comfort concept has occupied a prominent place in different ways, in both
indoor climate and working environment research. The optimisation of climatic
conditions in the indoor climate can prevent discomfort and illness and improve the
experience of comfort, while in the working environment, good comfort – objectively
speaking – contributes to increased and more concentrated work performance.

Fig. 3.6
THERMAL WELL-BEING
Baths in Vals, Switzerland, 1996.
Architect: Peter Zumthor.

In 1970, the Danish indoor climate researcher Ole Fanger defined the comfort concept by means of six measurable factors: air temperature, radiation temperature, air movement, relative humidity, metabolism or energy conversion and the thermal qualities of clothing. Subsequently, measures have been added for air pollution, quantity of light and limits related to the acoustic environment. Hence, indoor climate is a multi-dimensional play of complex inter-acting factors which include both natural and artificial conditions.

In his book *Design with Climate* (1963), Victor Olgyay graphically presented the interconnection between the traditional parameters of comfort. The two main axes are air temperature in Fahrenheit (the y-axis) and the air's relative humidity in per cent (the x-axis), respectively.

Above and below the comfort temperature, needs for wind and sunshine, respectively, occur. The purpose of the graph is to identify a comfort zone, in which there is no need for changes in temperature, humidity or air movement, and it clearly shows the tolerance to variations in the humidity and the need for a supply of radiation heat from the Sun if temperatures are below 70 °F or approx. 20 °C.

One might imagine that a spatial model would better illustrate the connection between the three most important parameters: temperature, humidity and wind. Each level in this model constitutes one of the parameters – heat, humidity and wind, and the levels each contain a comfort zone in relation to the specific climate parameter. If the three levels are put together, their common comfort zone would constitute a space reflecting the individual tolerance variation in relation to the individual parameters. This variation of tolerance is often described as a normal distribution curve (Gaussian curve), which in a spatial version takes on a spherical form as shown in Fig. 3.8.

On the basis of laboratory experiments, a series of specific standards or norms have been established for the comfort planning of indoor climate, which is embodied in ISO 7730 (4/26). In order to comply with these standardised comfort requirements it is necessary to use mechanical heating and cooling systems dimensioned to handle marginal situations and therefore over-dimensioned in relation to the normal situation. This leads to unnecessarily high energy consumption and the over-engineering of many buildings. The mechanically controlled systems also often cause discontent among the users who are unable to influence and adjust the climate to suit individual needs. This can lead to frustration and furthermore worsen the energy performance of the building.

The purpose of internationally standardised comfort requirements to the indoor climate is to achieve a uniform, measurable indoor climate with the same temperature in all rooms, and this is practically only possible through energy and resource demanding mechanical climate systems. However, ensuring a uniform climate and standardised comfort requirements free of unpleasant influences by mechanical means is often not sufficient to satisfy the user's need for well-being, variety and

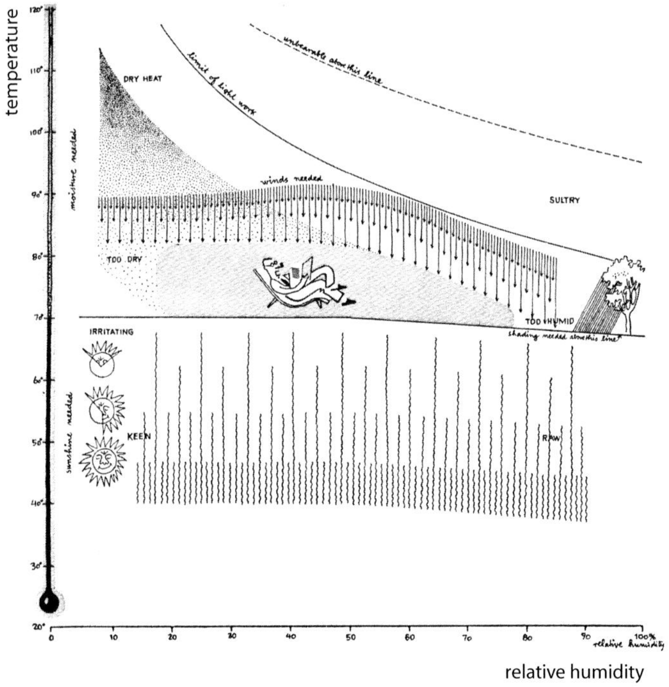

relative humidity

Fig. 3.7
COMFORT ZONES
From the book *Design with Climate*
(1963) by Victor Olgyay.

comfort conditions and the natural desire to create rooms of high indoor climatic quality based on natural conditions.

There is no way that traditional architecture can live up to these standardised and idealised comfort requirements; it does, however, offer a multitude of examples of how buildings and façades are designed to actively protect against, utilise and regulate different climatic conditions with minimum consumption of resources. Vernacular buildings offer more elementary experiences of space, matter and climate usually working with, rather than against, the principles of nature. Climatic elements such as light, sound, warmth, wind, humidity and precipitation affect us through the five senses: sight, hearing, touch, smell and taste. They are the basis for more holistic approaches to climatic design.

Insight into these principles and awareness of the significance of sensory perception to the experience of our immediate surroundings, are an important source of inspiration when designing spaces and architecture with a high level of comfort and utilising less resources. A more varied comfort concept shifts focus from the idea that climate variation and quantitative deviations from the norm imply a risk of physical discomfort to a more qualitative perception of man's need for a varied climate. It includes also the idea that climate variation can be stimulating and contain sensuous qualities that support the experience of architectural space, form and matter.

Therefore, the contemporary comfort concept argued here includes qualitative requirements for climatic variation, positively stimulating experiences and adaptation to individual needs, as well as greater quantitative consideration of energy consumption and the environment.

The variation of climatic experience both indoors and out is often an essential part of life – and to some it is the essence of being alive.
Oliver, P. *et. al*: *Encyclopedia of Vernacular Architecture of the World*, 1997.

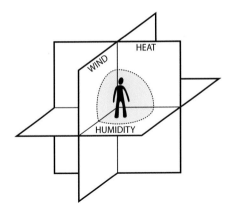

Fig. 3.8
THE COMFORT CUBE
The three fundamental parameters of climate gathered in a spatial model with the comfort zone as a sphere at the centre.

ADAPTATION AND CONTROL

Traditional Climate-adapted Architecture

Large Climate Screens

Traditional Climate-adapted Architecture

Georg Rotne
and Nanna Albjerg

I saw many huts that the natives made. They were all alike, and they all worked. There were no architects there. I came back with the impression of how clever was the man who solved the problems of sun, rain, wind.

(Louis Kahn, citeret efter en rejse til Afrika i 1961)

Fig. 4.1
CLIMATE ADAPTATION
A shaded square surrounded by blocks of heavy buildings, which ensures thermal stability in the buildings. Cairo, Egypt.

The origin of traditional architecture around the world is found in the elementary conditions of life. It is directly dependent on the intelligent use of nature's resources in any given climatic condition. From a resource and climatic point of view, examples of traditional building can be a rich source of inspiration for the complicated, high-tech, energy and resource demanding architecture of the 21st century.

An optimum solution to climatic design will almost always be a response to many different requirements, both measurable and immeasurable. It is also often culturally conditioned. For this reason, the best solutions cannot be created on the basis of optimisation inspired by individual climate parameters or individual, isolated comfort conditions. There will always be a context that can give a particular focus to a project, but a good overall solution presupposes the involvement and weighing of many, often conflicting, parameters.

Often, traditional architecture is not in keeping with the times functionally, technically or in terms of indoor climate when considered in relation to present-day requirements. However it facilitates insight into simple, comprehensible connections between man, space, built form, façades and the forces of nature. Climatic elements such as light, sound, warmth or chill, wind, humidity and precipitation connect with all of the body's senses. This activates the intellect and our relationship to basic sensuous qualities in our surroundings. Climate is an important starting point and an inexhaustible source of inspiration for architectural design. The interest today in living in the countryside may be seen as an expression of a fundamental need for elementary experiences, such as proximity to nature, the greater exposure to light and dark hours, the weather and the changing seasons.

Chosen examples

Examples of exemplary, culturally conditioned, climate-adapted architectural traditions can be found all over the world. In the following, three examples are described, taken from different cultures and climate zones, whose common denominator is that in each the design methods employed may serve as inspiration for contemporary, climate-adapted architecture.

The three examples chosen are:
- The traditional Arabian town house in Cairo located in a hot and dry desert climate.
- The traditional Japanese house in a humid, subtropical climate.
- The glass-covered town houses in La Coruña in northern Spain in a west European coastal climate.

Cairo

Climate control by means of ventilation and mass

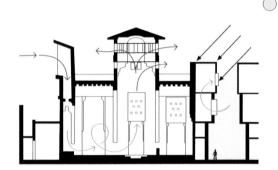

Egypt is situated in the world's largest desert, which stretches from the Atlantic Ocean across northern Africa and the Arabian Peninsula to the Persian Gulf.

All of Egypt's cultural societies have been located by the Nile, which turns a small strip of land into a fertile area on its long journey towards the north from the Sudan to the Mediterranean Sea. On both sides of the Nile Valley there is arid desert.

The Nile is also Egypt's most important transport route. Traditionally if you headed north, you followed the current. If heading south, you used a sail and the constant north wind which also cools and ventilates towns and buildings along the path of the Nile. The Islamic quarter in Cairo is laid out so as to utilise the north wind. The main streets running north–south lead the wind through the city, whereby the heat and stagnant air is drawn out of the narrow crossing alleys. Today, the quarter is in a severe state of decay, yet full of cultural landmarks and currently undergoing restoration.

The city is densely built with tall buildings featuring corbelling that provides cooling shade in the streets. The buildings constitute a huge mass of walls built together, which takes the edge off the temperature variations between the heat of the day and the cool of the night. The individual houses do not reveal their size or wealth outwardly, but have richly decorated courtyards and interiors hidden from view.

The entrance to the house is modest and designed to prevent straight views from the street through to the courtyard, which is in the centre of the house and the unofficial meeting place. The courtyard is shielded and shaded by buildings and only open towards the sky. It constitutes a reservoir of the night's coolness well into the day. In addition to the courtyard, large houses often have a walled garden where plants and water add to the feeling of a cool paradise.

Normally, the courtyard contains one or more covered transition zones which lead to the inner house.

Fig. 4.2
THE CLOSED EXTERIOR
The individual dwelling does not reveal its size outwardly. Access from alley to courtyard is broken to prevent people from looking in. Bayt Zaynab Hatun (15th century) restored and made accessible. The street level is raised due to centuries of accumulated rubbish.

Fig. 4.3
THE COURTYARD – THE CENTRE OF THE DWELLING
The house Al-Sinnari (18th century) was located next to a now filled-up canal with lush gardens. The house, which became home to Napoleon's artists and scientists, has been measured, described and restored. At the back, there is a spring, at the centre a fountain, to the right the raised *takhtabush* below the *maqa'ad*, the covered reception loggia.

The *takhtabush* is a covered living space, raised slightly above the level of the courtyard, cooled by the natural draught passing from the courtyard to the building.

The *maqa'ad* is a covered loggia, which serves both as a buffer zone and as a main entrance to the inner house. It may be situated above the entrance from the street to the courtyard, above the *takhtabush* or at ground level.

The most important internal space in the traditional Islamic house is the *qa'a*, which serves both as gathering place for the family and as official reception room for visitors. A house may have one or more *qa'as*, which may be situated at courtyard level or higher up in the building. The *qa'a* is tripartite with a *durqa'a* in the middle and an *iwan* on either side – a large *qa'a* faces north and a smaller one faces south. The *durqa'a* is a high room, stretching upwards through the building, ending in a lantern that provides light and airing.

The house Al-Sinnari (18th century) was located next to a now filled-up canal with lush

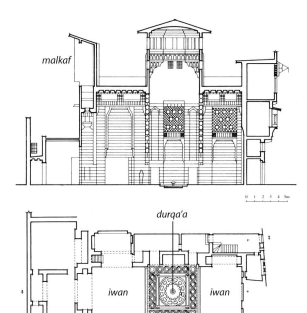

Fig. 4.4 a-b
SECTION AND PLAN
Qa'a with *durqa'a*, two *iwans* and *malkaf*. Al-Muwaqqi, built around 1350.

35

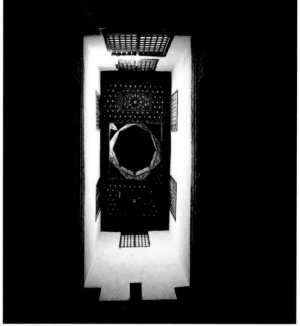

Fig. 4.5
OUTDOOR LIVING AND OVERVIEW
Above the entrance, the covered reception loggia, the *maqa'ad*, allows a view across the courtyard through the intermediate building to the garden. Manzil Suhaymi (17th and 18th century).

Fig. 4.6
LANTERN FOR LIGHT AND AIR
Above the *durqa'a*, the space continues in a lantern that provides light and natural ventilation. Manzil Suhaymi is the largest restored Islamic town house in Cairo featuring a house, a courtyard and a garden.

gardens. The house, which became home to Napoleon's artists and scientists, has been measured, described and restored. At the back, there is a spring, at the centre a fountain, to the right the raised *takhtabush* below the *maqa'ad*, the covered reception loggia.

Below the lantern, the space features a lowered marble floor and often a fountain. The floor in the two raised *iwans* is covered with carpets and cushions used for sitting or lying down on, and it is custom that no one sets foot on it wearing footwear.

The *qa'a* may contain a *malkaf* or wind tower, raised above the roof. The tower catches the breeze from the north and directs it down into the *qa'a*. A *malkaf* may be combined with a *salsabil*, an inclined marble plate with running water and carved wave-like curves that delay the water and encourage further evaporation and cooling.

The majority of the dwelling's light apertures, both towards the street and towards the courtyard, are screened by wooden boxes suspended from the façade or supported on corbels. The apertures are covered by wooden grills, *mashrabiyas*, which allow people to look out but not to look in. They are made of small, rounded pieces of wood. They filter light, provide privacy, discourage the entry of insects and, being made of flexible shutters, can be adjusted to external climatic conditions.

The traditional Islamic dwelling is a composition of outdoor spaces, buffer zones and indoor spaces, which individually and combined are designed to provide maximum comfort in an extreme climate. Often the spaces are multifunctional and used according to the climatic rhythms of the day and the year – a disposition of the dwelling that is completely different to present day western dwellings, in which the indoor climate is kept at a constant by means of mechanical energy supply, and where most rooms are confined to one particular function.

In the latter half of the 20th century, industrialisation and international modernism led to western-style housing being constructed in Egypt. Few opposed the importation of alien construction methods although throughout his professional life, the Egyptian architect Hassan Fathy (1900–1989) advocated the use of traditional materials, designs and climatic principles. He had his own home in the centre of the Islamic quarter in Cairo – a traditional dwelling, which is now being planned as a museum for the architect and his pioneering work. In his book *Architecture for the Poor* (1973) Fathy argued for a movement for climate-adapted architecture using traditional methods. In the book, he writes:

Today, we do not consider what we lose by not working with nature. However, if we consider earlier times' climatic solutions – such as wind catchers making allowance for wind direction and aerodynamics, and the marble plate with its carved waves, delaying the water's journey to the pool – we find a culture that has vanished with the introduction of the air conditioning system.

Fig. 4.7
RECEPTION AND LIVING AREA, *QA'A* AND *IWAN*
Lowered *durqa'a* with fountain and large *iwan* with *mashrabiya* on the 1st floor in Manzil Harawi (17th century).

Fig. 4.8
LIGHT, AIR AND VIEW OUT BUT NOT IN
Reception room on 1st floor ending in a large *iwan* with a *mashrabiya* above the entrance. Manzil al-Sinnari (18th century).

Fig. 4.9
NEW INTERPRETATION OF AN OLD THEME
At the Arab World Institute in Paris (1987), the architect Jean Nouvel designed the south-facing façade as a paraphrase of the classic *mashrabiya*.

Japan

Climate control via openness and flexibility

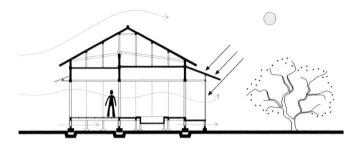

Fig. 4.10 a-b (opposite page)
TAKEYOSHI MIGITA HOUSE
Plan and section of detached house built during the
second half of the 19th century. From the main
room, *dei*, there is a gradual transition from inside to
outside across the building through the internal
veranda, *sotohara*, to the small external veranda
hi-en. The primary climate control is found between
the internal and the external veranda and consists of
three screens: two wooden doors shielding against
rain, *amado*, and a sliding door of latticework clad in
translucent paper, *shoji*.

Japan is an island kingdom consisting of four large
and thousands of small islands situated in the
Pacific Ocean off the Asian mainland. The landscape is rich in mountains and forests with volca-
noes and earthquakes. Settlement is concentrated
in the lowland towards the Pacific Ocean, with
one of the world's highest population densities.
The climate varies from cold temperate to tropical.
Most of Japan is situated in the subtropical zone
and has a mild, warm and humid climate. Spring
and autumn are short, and the summer is long,
warm and wet.

In the 14th century, a Japanese philosopher
wrote: *A house should be built with the summer
in view. In winter one can live anywhere, but a
poor dwelling in summer is unbearable.*

Thus, one great difference between a northern
European house and a Japanese one is that ours
protect against the winter and theirs against the
summer.

In Japan, the original religion is Shinto, according to which everything in nature is animated.
Buddhism is the other great religion that adds

simplicity and frugality to architecture and
landscape gardening. House and garden are the
simple, unadorned framework that makes it possible to appreciate the most minute detail in
everyday life. The most extreme simplification
is found in Zen Buddhism, as it is seen in the
garden at Ryoan-ji in Kyoto, where the sand
represents the ocean and the rocks the island
kingdom.

The traditional house is detached in a garden
or it may be a terraced house with a garden in the
city. House and garden are spatially integrated.
The garden is an abstraction of nature – a borro-
wed landscape.

The compositional principles of the Japanese
house are balance, asymmetry, immutability and
perspective. The sliding walls of the house permit
changing perspectives and limited views. Outside
the house, most often facing south, there is a
buffer zone, a covered balcony from which the
garden and the moon can be contemplated. The
buffer zone may be closed by means of a double
system of sliding doors and shutters. The house

Fig. 4.11
ZEN, MINIMISATION AND SIMPLIFICATION
In the courtyard by the Zen temple Ryoan-ji, the
Japanese island kingdom is symbolised by pieces
of rock placed in fine, raked sand.

Fig. 4.12
MODULES AND FLEXIBILITY
The traditionel Japanese dwelling divided into modules
with *tatami* mats. Room division and connection to the
outside are varied by means of light sliding walls.
Traditional country house, Hattoji.

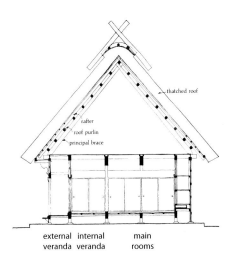

thatched roof

rafter
roof purlin
principal brace

external internal   main
veranda veranda   rooms

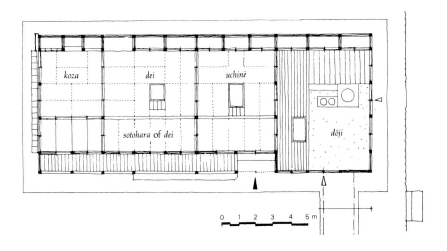

koza   *dei*   *uchinē*

sotohara of *dei*

dōji

0 1 2 3 4 5 m

Fig. 4.13
TRANSITION ZONE
The façade with screens and shutters facilitates
variation in the connection between inside and outside
and in relation to the rhythms of the day and the year.

Fig. 4.14
VARIABLE SCREENING
Traditional room in a recent dwelling in Shinjuku-ku, Tokyo.
Skoji sliding doors. The paper dims the light and reduces
draughts. Architect: Junichi Kawamura.

is raised above the terrain in order to separate
it from the moist ground and to ventilate the
floor. The roof's projection diverts the rain and
screens out the summer sun but allows the
low-angled winter sun access to the house.

The traditional Japanese house is slender,
flimsy and light. The structures have no
braces, and both on the inside and on the
outside, architecture is dominated by the
vertical and horizontal lines of the visible
structure. The materials, which are natural,
are often untreated. The house is constructed
for thorough ventilation beneath, through and
above the dwelling, and the walls are light,
movable screens. The house may be built in
one or two storeys, and the façade may
consist of a number of movable screens,
creating comfort during changing conditions
throughout the day and year. Curtains, blinds
and paper panels impede the view but allow
ventilation. Wooden shutters to close the
house during the night are stored in an
external cupboard during the day.

The Japanese house is standardised and
prefabricated, and the floor area is indicated
in the number of *tatami* mats. The mat, which
insulates against draughts from below, is
dimensioned according to human scale,
approx. 180 x 50 cm. It has a thickness of
between 45 and 60 mm, is packed hard with
rice straw and no one steps on it wearing footwear. The house has very few pieces of furniture, no beds and no chairs. You sit on the
floor, which means that eye height and thus
the horizon are lower than in the western
dwelling. The floor surface may be divided into
rooms by means of sliding walls, but most
rooms have no specific function, and a sleeping mattress can be rolled out anywhere
where there is a *tatami* mat. In this way, small
dwellings can house many residents without
losing spatiality. During the summer and during
daytime, the house is open, light and airy.
During the winter and at night, it can be closed
to the exterior and divided into smaller
rooms.

Japan was not opened to the West until
the end of the 19th century. Japanese culture
quickly became an inspiration within painting
as well as graphic design and modern architecture. Mies van der Rohe's Barcelona Pavilion
has a number of features in common with the
traditional Japanese house. An open layout,
undefined transitions between outside and
inside, controlled experiences of courtyard
spaces, free-standing walls and an eye level
halfway between floor and ceiling.

Fig. 4.15
MAXIMUM VENTILATION
Teahouse constructed of untreated natural materials and raised above ground with ventilation beneath, above and through the house. Projection screening out the summer sun and allowing the low-angled winter sun to enter, whilst at the same time providing some protection against rain. Terrace for meditative contemplation of the garden and the moon.

Fig. 4.16
MODERNISM'S INHERITANCE FROM JAPAN
Mies van der Rohe's Barcelona Pavilion from 1929, pulled down and recreated in 1986, reinterprets simplicity, openness and refinement in the Japanese house.

Fig. 4.17
DIVISION INTO ZONES
Several layers of screens form transition zones between the inner house and the garden. Nomura House, Kanazawa.

La Coruña

Glass galleries as climatic buffer zone

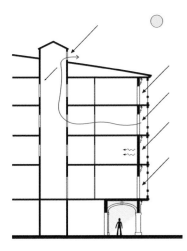

In modern day experiments with natural climate control, one of the most popular solutions is a double façade with an intermediate buffer zone. The solution is not new. It was prevalent in many places across Europe towards the end of the 19th century, when industrialisation led to cheap glass in large quantities. The largest concentration of glass-covered façades is found in La Coruña in the north-western corner of Spain.

La Coruña is situated by the Atlantic Ocean on the Bay of Biscay. The city was founded on a peninsula, which was easy to defend, and it has a natural harbour. Due to its location, La Coruña became one of Spain's most important seaports, opening the country towards northern Europe and America. 'The invincible armada' sailed from here heading for England, where it was defeated by storm and the British in 1588.

In the course of the 19th century, the town grew due to industry and trade, spreading across the narrow isthmus towards the mainland. Today, the isthmus features a dense city quarter between a large, sandy beach and an active port. It is in this quarter that most of the glass galleried houses are found. The waterfront, which was constructed during the second half of the 19th century, forms a cohesive glass wall, which has given La Coruña the nickname *The Crystal City*.

The climate in La Coruña is similar to that of Southern England and Northern France, with many hours of sunshine, but with plenty of rain, frequent wind and occasional storms. The glass galleries form a climatic buffer zone on the outside of the stone façades, where they shield against wind, rain and cold. The glass façade induces passive solar heat, protects against noise and provides the residents with the possibility of observing life in the street under comfortable conditions.

The galleries owe their inspiration to sailing ships which had a tradition for converting open balconies into protected glazed living areas without obscuring the view. The sailing ships

Fig. 4.18
THE CRYSTAL CITY
A cohesive glass wall meets the
protected port of La Coruña.

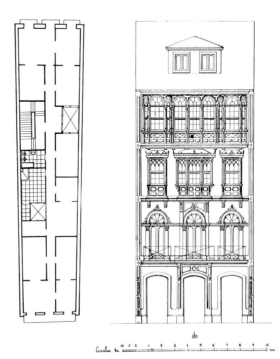

Fig. 4.19 a-b
NARROW AND DEEP DWELLINGS
Plan and façade of glass galleried building by the
waterfront. The galleries create a double glazed
façade on the upper floors, which contribute passive
solar heat to the rooms behind. Vertical shafts inside
the buildings ensure daylight and ventilation.

43

Fig. 4.20
UNITY AND COHESION
On three sides, the large, square Plaza de María
Pita has continuous façades with arcades at the
bottom and glass galleries at the top.

of the baroque era had aft cabins, which became models for La Coruña's glass galleries. The construction of the galleries continued into the 20th century and then died out under the influence of international modernism and a changed social pattern.

Two connected building complexes with glass galleries stand out from the rest. One of them is the Plaza de María Pita, the other is Avenida de la Marina at the waterfront.

The square Plaza de María Pita was constructed in 1875 on the edge of the old town, standing back from the harbour. The square is formed by the city hall and a U-shaped complex of 25 identical buildings with continuous glass galleries on the upper floors. The individual buildings are divided vertically by arcade apertures and glass galleries and horizontally by three zones: at street level, a dark arcade set back somewhat; at the centre, two traditional floors, and at the top, a glass gallery. The divisions add harmony and weight to the square. The buildings along the waterfront at Avenida de la Marina, were constructed between 1869 and 1884. The houses are four–five storeys high and between three and five bays wide. At ground level there are shops; the 1st floor usually contains offices, and above these there are dwellings. Continuous glass galleries cover the first to fourth and fifth floors and face towards the sun and the harbour. The sites are long and narrow, and the buildings have glass galleries towards the waterfront and more dispersed glass-covered balconies in the parallel back street. Although the glass façades are connected, each individual gallery's characteristics and different details add a variation to the whole, so that the façade of waterfront neither appears monotonous nor chaotic. The building style is both classical and expressive of the environmental principles employed.

Under the influence of international modernism, glass façades were built throughout the 20th century without consideration for passive solar heat gain and the consequent need for indoor mechanical climate control – normally through air conditioning. Towards the end of the century – and influenced by fuel crisis and ecological awareness – traditional glass galleries did, however, become models for a new generation of glass façades.

Fig. 4.21
CLIMATE PROTECTION
The dense city that grew up on the isthmus between beach and port towards the end of the 19th century, features a rich and varied selection of glass galleries.

Fig. 4.22
NEW INTERPRETATION OF GLASS GALLERIES
At La Coruña's waterfront, this building demonstrates a new interpretation of the glass gallery theme.

Large Climate Screens
Glass-covered spaces – from the Crystal Palace to the Eden Project

Georg Rotne

The scientific inventions of the Enlightenment, the technical advances of the Industrial Revolution, and the ever increasing travel activity towards the end of the 18th and the beginning of the 19th century, spawned new ideas and visions about both climate adaptation and house construction. Industrialisation led to a reduction of the price of glass and iron, which were used to build the large transparent structures for the new building types that emerged in connection with growing transport, trade and urban growth: greenhouses, railway stations, glass covered shopping passages and exhibition pavilions etc.

Two pioneers within the development of greenhouses, the Scotsman John Claudius Loudon (1783–1843) and the Englishman Joseph Paxton (1803–1865), worked both as gardeners and as architects. Loudon was one of the 19th century's most productive writers in the field of horticulture and architecture. Both his writings and his experimental glass houses were far ahead of his time. In his *An Encyclopaedia of Gardening* (1822) Loudon writes:

Indeed, there is hardly any limit to the extent to which this sort of light roof might be carried; several acres, even a whole country residence, where the extent was moderate, might be covered in this way, by the use of hollow cast-iron columns as props, which might serve also as conduits for the water which fell on the roof. Internal showers might be produced in Loddiges' manner; or the roof might be of the polyprosopic kind, and opened at pleasure to admit the natural rain. Any required temperature might be kept up by the use of concealed tubes of steam, and regulated by the apparatus of Kewley. Ventilation also would be effected by the same machine. The plan of such a roof might either be flat ridges running north and south, or octagonal of hexagonal cones, with a supporting column at each angle, raises to the height of a hundred or a hundred and fifty feet from the ground, to admit of the tallest oriental trees, and the undisturbed flight of appropriate birds among their branches. A variety of oriental birds, and monkeys, and other animals might be introduced; and in ponds, a stream made to run by machinery, and also in salt lakes, fishes, polypi, corals, and other productions of fresh or sea water might be cultivated or kept. The great majority of readers will no doubt consider these ideas as sufficiently extravagant; but there is no limit to human improvement, and few things afford a greater proof of it than the comforts and luxuries man receives from the use of glass …

Paxton became known for his greenhouses and acquired fame because of the Crystal Palace constructed for the World Exposition in London in 1851. Crystal Palace was the first example of a large (75,000 m²) industrially manufactured building. Simple, rational, prefabricated and joined in six months in Hyde Park it has remained an inspiration for architects and engineers ever since.

Later on the Crystal Palace was taken down and reconstructed in an enlarged version in Sydenham, South London, where it was destroyed by fire in 1936.

Fig. 5.1
PALM HOUSE
Bicton Park Botanical Gardens, Devon.
Inspired by Loudon. Built around 1825–30 by
Lord John Rolle for his wife, Lady Louisa.

Fig. 5.2
PALM HOUSE
The Palm House at Kew Gardens, London, 1844–1848.
Architect: Decimus Burton.
Engineer: Richard Turner.

Fig. 5.3
THE GLASS COVERS OF THE CITY
Galleria Vittorio Emanuele II, Milan, 1865–1877.
Architect: Giuseppe Mengoni.

Whereas nothing is left of the Crystal Palace, the contemporary Palm House at Kew Gardens is well preserved. As one of the most beautiful examples of 19th century greenhouses, it unites classical architecture with industrial materials and technique and is a rare, harmonious result of collaboration between architect Decimus Burton (1800–1881) and engineer Richard Turner (1798–1881). In terms of beauty, the Palm House at Kew Gardens was not surpassed by any of the palm houses built during the rest of the century to provide the middle classes with the opportunity to experience the tropical nature of the colonies.

The culmination of exotic phantasmagoria was reached at the end of the century, when the Belgian King Leopold II between 1865 and 1899 had the architect Alphonse Balat construct an extensive complex of glass houses and glass galleries in the Royal Gardens in Laeken outside Brussels. This building complex constructed between 1865 and 1899 allowed the king, his family and guests to wander several kilometres in a tropical climate under glass.

Leopold II founded the Congo Free State, and regardless of its name, he considered this his private property, which brought him inconceivable wealth through heavy-handed plundering. Many years of criticism would pass before Belgium in 1908 nationalised the colony and changed its name to Belgian Congo. The following year, Leopold II died in one of his glass houses – in the midst of the artificial world he considered his oasis from the demands and realities of life.

Whereas the greenhouses create an artificial climate for a transplanted nature, the glass-covered pedestrian passages provide comfort for small sections of the city. The first passages were built in Paris at the beginning of the 19th century. Throughout the century, they grew both in number and in size, until in 1877, they culminated in the Galleria Vittorio Emanuele II in Milan. This passage remains well preserved and functional due to its central location. The passage, which was constructed in celebration of Italy's unification, appears eclectic in style, highly decorated and with coloured glass in the pendentives of the dome.

At the beginning of the 20th century, glass was celebrated by German expressionists, whose poetic leader, Paul Scheerbart (1863-1915) hated bricks and felt a fanatical love for glass as a symbol of life, openness and joy, as expressed in the following mottos in his lertter to Bruno Taut (1914): *Glass opens up a new age / Brick building only does harm / Bricks may crumble / Coloured glass endures. / Greater than the diamond / Is the double walled glass house.*

In his visions, Scheerbart foresaw the double glass structure as a means of controlling heat loss, overheating and ventilation, and that a new material, transparent and elastic, would emerge and outdo glass.

The group Gläserne Kette formed around Scheerbart shortly before his death. The architect Bruno Taut (1880–1938) drew and wrote the book *Alpine Architektur – Eine Utopi*, in which the mountain tops are crowned by crystalline buildings, and for the Werkbund Exhibition in Cologne in 1918, he constructed a pavilion of multi-shaped and multicoloured glass, which he dedicated to Scheerbart. The movement

Fig. 5.4
THE ROYAL GARDENS IN LAEKEN
From the construction of the winter garden
at the Royal Gardens outside Brussels.
Large climate screens.

Fig. 5.5 a-b
INSPIRATION FROM THE COLONIES
The large greenhouses of the 19th century
created an artificial climate for nature from
the tropical colonies. Palm houses in the
Royal Gardens at Laeken, Brussels
(1874–1895).
Architect: Alphonse Balat.

Fig. 5.6
COLOURED GLASS
Wayfarers Chapel, 1951.
Architect: Lloyd Wright.

Fig. 5.7
DOME OVER MANHATTAN
Buckminster Fuller's conceptual project for a
geodetic dome over Manhattan, 1960.

did not have many buildings to show. One successor is Lloyd Wright's Wayfarers Chapel by the Pacific Ocean at Palos Verdes (1951). For the rest of the century, glass was mainly transparent and used rationally.

Electricity, growing mechanisation and a belief that man could be master of nature led to closure of the climate screen and mechanical control of indoor comfort. At the same time, as a result of a change in building technology, the climate screen became independent of the load-bearing structure. From the middle of the 20th century, the high-rise building became a symbol of modernity and capital across the globe. A box on edge with thin, closed sides of glass and with installations for mechanically controlling climate placed at the top.

The vision of covering large areas with glass and creating an artificial climate was carried on into modernism's rationality and technical possibilities. The American inventor Richard Buckminster Fuller (1895–1983), who developed geodetic domes and built the USA's pavilion for the World Exposition in Montreal in 1967, elaborated a project for covering part of Manhattan with a dome with a 3 km diameter. The German architect Frei Otto (1925–), who constructed the German pavilion for the same World Exposition, prepared a project for a city in Antarctica covered by a translucent, net-reinforced double membrane.

Neither Buckminster Fuller's large dome nor Frei Otto's Arctic city were however realised, but in shopping centres, office atria and museums alike, indoor landscapes with controlled climate and transparency have become a reality.

However, the vision of a city under glass lives on. In Herne, in the Ruhr district in Germany, the French architects Jourda and Perraudin constructed a hangar of wood and glass in 1999 to favour a town and conference centre with a subtropical climate. In a former clay pit in Cornwall in the UK, the architect Nicholas Grimshaw constructed the ecological experience centre the Eden Project in 2001 as a number of inter-connected domes constructed with a lightweight steel frame filled with hexagonal cladding panels of three layers of inflated ETFE plastic. The project is the world's largest greenhouse facility. Both the project in Germany and in the UK have the climate screen as an independent, enclosing structure, which covers and protects a mini world with an artificial climate – the first with and the second without climate control.

Fig. 5.8
PLASTIC MEMBRANE
Development within plastic materials in particular has
created new possibilities for creatiing huge, transparent
coverings. ETFE plastic membrane for the Eden Project
in Cornwall, 2001.
Architect: Nicholas Grimshaw & Partners.

CLIMATE THEMES

Hot and Cold

Humidity and Precipitation

Wind and Ventilation

Light and Shadow

Climate Themes

The following chapters discuss the most important climate parameters: heat, humidity, wind and light, and how these interact with man and building. Obviously, it is difficult to describe these parameters individually. They appear in both the outdoor and the indoor climate in a continual interplay. However, detailed knowledge of how we and the building react to the individual climatic influence can open up for strong architectural detail solutions, which functionally as well as aesthetically can give the building a richer expression. The individual chapters describe the climatic theme on the basis of how it appears in outdoor and indoor climate, and how it affects human comfort. Similarly, it is described how the building is designed traditionally and currently to utilise, adapt to or protect it against climatic influences, and which behavioural and technical options we have for controlling and optimising both comfort and consumption.

Heat   Temperature is a central factor in the experience of both outdoor and indoor climate. Thermal comfort affects the dimensioning when planning indoor climate. However, the temperature is highly influenced by two other important climate parameters: humidity and air movement/the wind. Both parameters affect the thermal level and our experience of hot and cold. The balance between heat addition and heat loss plays a significant role in relation to man's comfort, the building's energy consumption and the climate of the world.

Humidity   The humidity of outdoor climate in the form of rain or snow is not only a tangible climatic influence, but also in an architectural context a challenge and inspiration. The design and expression of the façade often reflects – at times notably – efforts to protect against precipitation and drainage of water. The patina caused over time by water along with other climatic influences also contributes to the architectural expression.

Wind   The wind acts as both friend and foe – as a cooling breeze, appreciated in the summer heat, or as biting cold during winter storms. Buildings' orientation and design can reduce the wind's negative effects and further the positive effects, e.g. by creating ventilation and keeping the structure sound.

Light   Light is a prerequisite for a healthy, comfortable environment rich in experiences. In architectural terms, the challenge is to capture the character of the local light, designing apertures and surfaces so that light is utilised optimally functionally and aesthetically, and to lower and enhance according to need and desire.

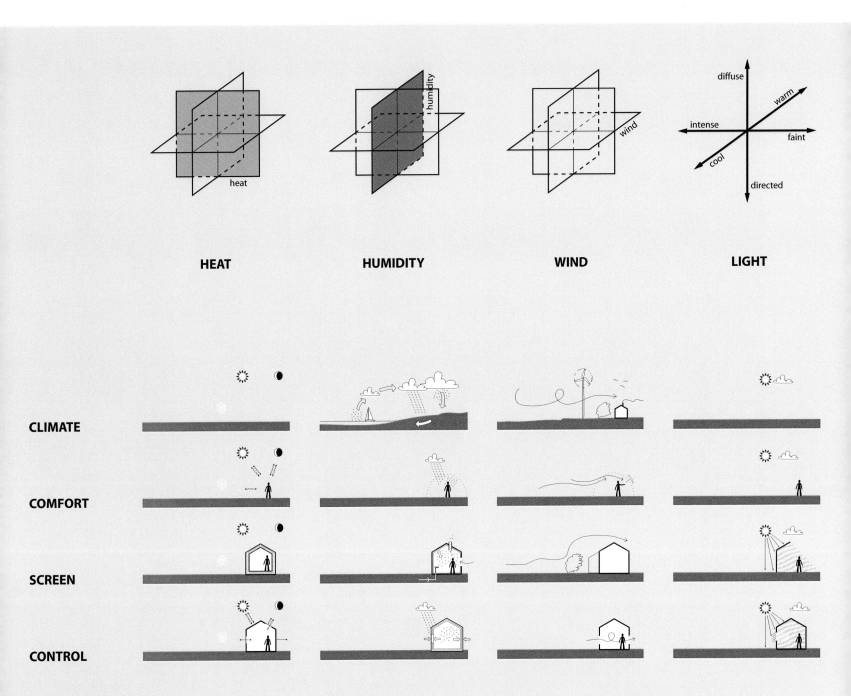

HEAT **HUMIDITY** **WIND** **LIGHT**

CLIMATE

COMFORT

SCREEN

CONTROL

Hot and Cold

Torben Dahl
and Eva Tind Kristensen

The term 'heat balance' has gained global topicality. The world's heat balance is unstable and it has been established that the Sun supplies more energy than the Earth emits resulting in rising temperatures and global warming. The balance between supply and emission of heat is also decisive at a smaller scale – at the level of the body and the building.

Both indoor and outdoor climatic conditions are to a great extent crucial to our health and personal energy levels. It is a known fact that you lose energy and initiative if the climatic conditions – and in particular the temperature – deviates a lot from the comfort level for a prolonged period of time. Man is a warm-blooded being who needs a constant body temperature of approx. 37 °C. The air temperature we prefer to be surrounded by is somewhere between 18 and 26 °C. However, air temperature is only one of many variables that affect our thermal well-being. This goes for both the body and the building.

THE BODY'S HEAT BALANCE

The body's ability to maintain the thermal equilibrium around 37 °C is a fine balance between a number of internal and external conditions, the most important being the body's assimilation of food (metabolism), the body's activity level and evaporation from the skin. The external conditions apart from air temperature are the radiation balance in relation to the surroundings, the body's exchange of heat with the surroundings by conduction and the air's humidity and movement. There are other factors that influence our well-being particularly at a psychosocial level, but in terms of thermal comfort, the above mentioned are the most important.

In order to consider the things that either supply us with heat or drain heat from us, the balance can be listed as follows:

Heat contribution
– Conversion of food (metabolism)
– Movement at micro level (goose bumps, shivers), medium level (huddling up, rubbing your shoulder vigorously to get warm), macro level (work, exercise, sports)
– Radiation towards the body
– Conduction from the surroundings
– Convection (in).

Heat emission
– Radiation from the body
– Conduction towards the surroundings
– Convection (out)
– Evaporation.

Fig. 6.1
CLIMATE SCREEN
Nuuk Culture House, Greenland, 1997.
Architect: schmidt, hammer lassen architects.

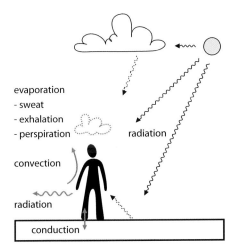

evaporation
- sweat
- exhalation
- perspiration   radiation

convection

radiation

conduction

Fig. 6.2
**THE BODY'S HEAT EXCHANGE WITH
THE SURROUNDINGS**

The body is in a constant thermal interchange with the surroundings both outside and inside when it is protected by the building's climate screen. It is estimated that in a normal situation within the thermal comfort zone, 40% of the body's heat emission takes place by radiation, 40% by convection and 20% by evaporation. This ratio is highly dependent on the surroundings around the person, and varies from one climate to another. Hence the balance for human comfort is not constant but varies. The role of architectural design is to strike the right balance and to provide user flexibility to support the best conditions for that location according to personal need.

In *Temperature and Human Life* by Winslow & Herrington (1949. Source: McCornick, *Human Factors Engineering*, 1964) it is shown that the relationship between the three types of heat emission is strongly affected by the temperature of the air and that of the surrounding walls. The illustration 6.4 shows the distribution of the body's heat emission at evaporation, radiation and convection, respectively, in different thermal situations with different surface and air temperatures.

Thermal well-being is essential for the experience of comfort. As man is a warm-blooded mammal, the thermal conditions of the surroundings should neither be too hot nor too cold. But the limits are wide. Man is capable of adapting to a dry and warm climate with average temperatures of 30–40°C, and likewise, the Arctic zones with temperatures as low as -50°C have been inhabited for centuries.

Clothing

Clothing provides a quick and immediate thermal control factor. Clothing creates an insulating layer around the body, which can reduce heat emission by conduction, radiation or convection. Clothing can also be used to reduce heat intake. White, reflecting material reduces the incoming radiation, and thick layers of material prevent heating by conduction, if the air is hotter than the skin temperature.

In recent years, leisure and sportswear has developed into a high-tech science, more based on defined functional requirements (functional clothing) than on empirically gained experience. Today, it is possible to buy clothes that correspond to the individual functions in the building's climate screen. Closest to the body there is a sweat-transporting layer, which is gentle to the touch, then follows an insulating layer, and on the outside, a rainproof layer also allows body moisture to penetrate. In some ways the façade or climate screen of a building functions in much the same fashion.

THE BUILDING'S HEAT BALANCE

In the world of architecture and building physics, 'the third skin' has been known to be used as an allegory for the building's climate screen and its climate controlling function. The climate screen is the overall external instrument for fulfilment of our comfort needs. It includes the walls, roof, floor and all the openings (windows, vents and doors).

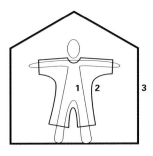

Fig. 6.3
THE THREE SKINS
1   The body's skin
2   Clothing
3   The building's climate screen.

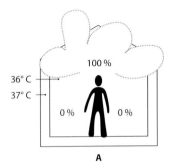
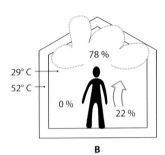
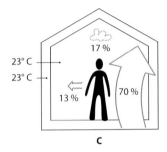
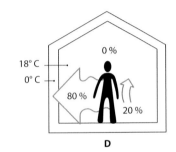

A

B

C

D

Fig. 6.4
THE BODY'S HEAT EMISSION

The illustration shows the body's relative heat emission at different air and wall temperatures (based on Winslow & Herrington, *Temperature and Human Life*, 1949).

In situation A, which corresponds to a day during a heat wave, the heat loss is almost entirely due to evaporation.

In B, where the walls are extremely hot and the air is kept cool (maybe by air conditioning), the heat loss is still dominated by evaporation.

In C, which corresponds to a normal North European summer day, the heat loss is more composite, convection being dominant.

In D, which is this book's authors' estimated extrapolation based on the source mentioned above, and which simulates a cold winter day, the heat loss is estimated to be mainly by radiation.

 evaporation   radiation   convection

Fig. 6.5
THE SPACE SUIT

The space suit is the heaviest and most advanced dress in the world. It costs €10 million, weighs 153 kg and ensures that the body is able to survive and work in extreme climatic conditions.

Fig. 6.6
COLD AS PLEASURE

Two East Greenland women in front of a tent, 1898. Clothing can protect, but the body is able to adapt to or even make it possible to enjoy extreme temperatures.

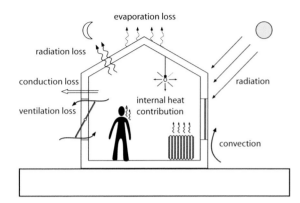

Fig. 6.7
THE BUILDING'S HEAT BALANCE

Fig. 6.8
INCIDENT SUNLIGHT
Wine bar in Ravello, Italy, 1890. P. S.
Krøyer. Heat in the sunlight on the floor
and coolness in the room's shadow.

The climate screen task is primarily to shield us from adverse external climatic influences, but also to absorb and utilise the influences that may improve comfort. It is the double function (protection and utilisation) which makes the building envelope such a challenge for architects. As is the case with the body's heat balance it is possible to analyse the building's heat balance. The building primarily receives its heat contribution from solar heat, the Earth's heat (geothermal and radiant) and the house's own heat supply (heating system, electricity consumption and people contribution), and the house emits its heat via heat conduction, radiation, convection (ventilation) and evaporation.

The building's heat contribution
Solar heat
Solar radiation is transferred directly to the building either through the building's apertures, where it is absorbed by the building's internal surfaces, or via the building's external surfaces, which absorb and transfer heat through conduction to the house's internal surfaces. The building can also be supplied with heat indirectly from the Sun's heating of the outdoor air, which through ventilation transfers heat to the building. The effect of the Sun's radiation varies according to the altitude of the Sun, the orientation and surface properties of the building and local climatic variations.

Earth heat
The relatively constant temperature below the Earth's surface (called geothermal energy) can be exploited both for heating and for cooling depending on need. During cold periods with a need for heating, the heat in the Earth's upper layer is utilised via heat exchange for heating of small units – typically one-family houses on large plots of land. The higher heat in the Earth's deeper layers can be exploited directly by pumping hot subsoil water up for heating and sending cooled water back into the ground. Geothermal energy systems are normally either surface systems (low level heat) or deep bore systems (high level heat).

The temperature rises 20–40°C per km from the Earth's crust down towards the core. A geothermal heating system that collects water from deep drillings is expensive and dependent on a large uptake area. Consequently, such systems are often established in urban areas that already have a water-based heating system, e.g. district heating. In connection with an existing system, geothermal heating is both a financially viable and resource-saving heating system. Since the temperature below ground is relatively constant and in summertime often more cold than the ambient air, shallow geothermal systems can be used for cooling as well as heating and can utilise air as medium as well as water.

The building's own heat production
In a normally functioning building, heat will be contributed from certain processes, partly from consumption of energy (lighting, cooking and other electrical appliances)

and partly from people who stay and work in the building. Where the comfort level cannot be satisfied by the energy derived from these processes and from the surroundings – Sun and Earth – heat must be supplied to the building from technical systems such as district heating, central heating systems, fireplaces, stoves and mobile ovens.

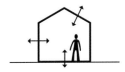

The building's heat emission
Just as with the body, the building also exchanges heat with the surroundings through conduction, convection/ventilation, radiation and evaporation.

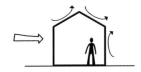

Heat conduction
In poorly insulated buildings, conduction is the primary reason for heat loss. Heat is conducted through the building's outer walls and roof and lost to the surrounding air. The heat loss is directly visible if during snowy weather the roof is clear of snow while the ground around the house is covered in snow. Insulation is needed to protect the structure against heat conduction, whether the objective is to keep the heat in or out. Insulation is normally achieved by placing a light, air filled material in the outer wall and roof structure and one or two air layers in window panes in windows and doors. Legislative regulations stipulate requirements to the thickness or U–value of the insulating layer.

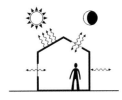

Convection
Air movements around and through the building reinforce the cooling of the building, both when this is desired during severe heat, and when it is less appropriate in cold weather, where the wind will add to the building's heat loss. The section *Wind and ventilation* discusses the significance of convection in greater detail.

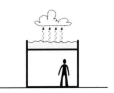

Fig. 6.9
THE BUILDING'S HEAT EMISSION

Heat radiation
A not insignificant part of the house's heat exchange with the surroundings takes place – as is the case for the human body – via radiation. The transfer of heat by radiation between two bodies depends on the surface temperature of the emitting (relative heat) and absorbing (relative cold) surface and the particular characteristics of these surfaces: emittance and absorptance. Radiation that hits a surface will either be reflected or absorbed, expressed by the coefficients reflectance (r) and absorptance (a). The sum of the two coefficients (r + a) for a surface will always = 1.

Light and shiny surfaces normally have higher reflectance than dark and matt ones. The perfect absorber – the theoretical black body – has a reflectance of zero. The black, cloudless night sky is close to the black body. Therefore, the building's heat loss is greater from the roof surface than from the walls, and this is why there are normally greater requirements to the insulation of the roof surface.

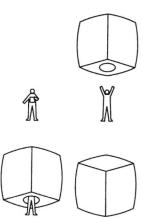

Fig. 6.10 a-b
BASIC HOUSE
In his transportable Basic House, Martin Ruiz de Azúa has utilised the knowledge about reflection and absorption to create a super light transportable shelter, which according to the architect can protect against both cold and heat simply by being turned over. The house is a cube consisting of a double metal-clad polyester cloth with golden cladding on one side and silvery cladding on the other. The Basic House is now part of the permanent exhibition in MoMA, New York.

Good examples of utilisation of knowledge of radiation may be found in e.g. the Portuguese cladding of houses with glazed tiles, often in light colours, and in the white walls of Spanish and Greek villages. In both examples, the strong solar radiation is reflected, and the mass of the outer wall is not heated as much as it would be otherwise.

In the planning of a building, heat intake and heat emission by radiation can be included in the calculation of a building's heat balance. This can be done partly by including the positive energy contribution to the building from incident sunlight through window apertures, partly by including the energy glazing's ability to reduce radiation loss through the glass.

Evaporation
The cooling effect of evaporation is known from the body's secretion of sweat, which occurs under high external heat influences or as a result of heat production at hard physical labour or illness. The sweat that settles on the skin's surface evaporates and thus uses heat, which is taken for the surface (the skin) where the evaporation takes place and hence cooling.

In construction, this relation can be utilised in warm climates to cool the building, e.g. by water being atomised and evaporating in the room, thereby lowering the air temperature, or by water flowing or being sprayed over the building's external surfaces, evaporating and cooling the building part and thus lowering the temperature in the building's room. Fountains in hot climates exploit the effect and are commonly found in domestic courtyards.

THE BUILDING'S HEAT CONTROL
The balance between a building's heat contribution and its heat emission may be influenced in the entire disposition of and material choice for the building, and the design of the façade systems which plays a decisive role in the control of the building's heat balance.

Heavy – light
The thermal reaction in relation to an external climate differs depending on whether a house has a large or small mass. Heavy materials conduct heat better than light materials, but some heavy materials such as stone also have great heat accumulating capacity or specific heat capacity. This gives great inertia to temperature changes. This inertia can be exploited in areas with great differences between day and night temperatures. High daytime temperatures heat up a thick and heavy stone wall. The wall 'stores' and emits this heat throughout the rest of the day. This is often called thermal capacity and is often exploited in today's buildings in the form of large areas of exposed concrete construction.

Conversely, the night-time cooling of the stone mass can keep the indoor climate pleasantly cool throughout the day. This means that a building's mass is used both

Fig. 6.11
COOLING BY EVAPORATION
This restaurant is situated in Nicosia, Cyprus; it completely encircles
its outdoor serving area with water atomisers. As the water
evaporates, the temperature at the tables drops tangibly.

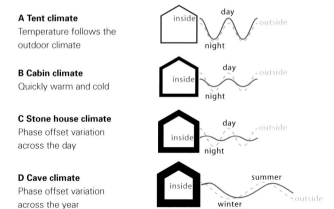

A Tent climate
Temperature follows the
outdoor climate

B Cabin climate
Quickly warm and cold

C Stone house climate
Phase offset variation
across the day

D Cave climate
Phase offset variation
across the year

Fig. 6.12
FROM LIGHT TO HEAVY

to even out temperature variations and to protect against extreme cold and heat, thus stabilising the indoor climate. As a result, buildings with great inertia and great weight are less sensitive to temperature variations in the outdoor climate than light buildings whose indoor climate is a direct reflection of the outdoor climate.

Generally speaking there are four climate architecture types – tent, cabin, stone house and cave.

The tent climate is related to the light façades of thin membranes such as flimsy glass, cloth and thin plastic materials. Facades built of these materials have poor insulation capacity, which makes the indoor climate directly dependent on the outdoor climate. Consequently, temperature variations will be felt immediately after a change of weather or during the day. So such buildings warm up quickly in the morning and cool down quickly after the Sun sets.

If the thickness of the façade is increased, the result is a *cabin climate*. The façades are still relatively flimsy – maybe they consist of a combination of boards, glass, layers of cloth or plastic separated by an air space. The house's indoor climate remains dependent on the outdoor climate, but a greater delay applies. Even relatively poor insulation and air spaces delays the temperature variation transfer to the indoor climate by several hours.

If the mass is increased further, you may refer to a *stone house climate*, a term that covers houses built of uninsulated natural stone, brick or concrete. The outdoor climate will now be reflected in the indoor climate throughout the day.

The cave climate creates an almost temperature-stable indoor climate without the use of insulation and it requires a very large mass. There are plenty of historical examples. Churches cut out from solid rock in East Africa. Subterranean town and building complexes in soft sandstone areas in the Mediterranean countries. In these places, the indoor climate shows little variation. The deeper into the ground, the less variation, and the difference in the indoor house temperature will vary over an entire year. Examples exist of contemporary archive buildings for humidity and temperature-sensitive papers and books that use great mass and thick walls combined with insulation to ensure a constant humidity and heat level.

The principle of using the thermal mass can be experienced today as it could a thousand years ago, and it can be studied in both past and present building structures. It is increasingly relevant today as society faces an energy crisis. In modern day sustainable construction, the heavy mass is used as a thermal regulator. Here it is often combined with lightweight high-tech materials which provide the benefit of tent climate to create architecture which is flexible and comfortable.

In the natural world earth is used as an environmental regulator. For example, African termites create highly advanced climatic structures both above and below ground.

The termite settlement has been a field of research for a long time among architects, biologists and engineers interested in energy and resource-saving

Fig. 6.13
STONE HOUSE CLIMATE
This French stone house's inertia to temperature changes stores the night's coolness for the day and the day's heat for the night – phase offset variation across the day.

Fig. 6.14
TENT CLIMATE
In this leisure cabin in Espoo, Finland, the indoor temperature is a quick reflection of the outdoor temperature.
Architect: Kaakko Laine Liimatainen Tirkkonen, 1992.

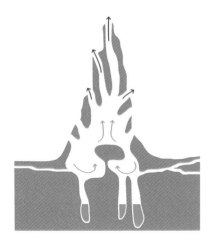

Fig. 6.15
TERMITE MOUND
Heating of the termite mound's upper part
draws chilled air up from the cool
subterranean part.

Fig. 6.16
EASTGATE DEVELOPMENT, ZIMBABWE
The building is cooled in accordance with the
principles of the termite mound.
Architect: Mick Pearce, 1996.

construction, because the termite building structure constitutes a well-developed natural ventilation system in which temperature can be controlled by means of mass and wind.

In the Eastgate Development project in Harare, Zimbabwe (1996), this knowledge is transferred to architecture. The building's ventilation system is developed on the basis of the thermal logic of the termite mound. The air in the roof space is heated by the Sun, creating thermal lift that draws cooling air into the lower part of the building, thereby reducing the demand for energy consuming air-conditioning.

At the Öijared Golf Club (1986) in Lerum, Sweden, architect Gert Wingårdh has designed the building to 'grow' out of the rock. A landscape of rock, earth and grass spreads across the building, giving it thermal stability. The building is intended as a part of the landscape, which is stressed by the fact that the roof can be walked upon, and that at the same time it constitutes the teeing ground for the 1st hole on the golf course.

Zone division

The combination of heavy and light materials can be optimised by dividing the building into thermal zones with different choices of material and functionality depending on the seasons. This can be done both horizontally and vertically. The classical Danish master builder's house is an example of the division of a building into climate zones from cellar to attic, where each floor is of a structure made of different materials, resulting in different temperatures and atmospheres. This division creates a greater physical stimulant than a completely homogenous indoor climate.

In earlier days, it was common for each room in a dwelling to have its own temperature: a warm kitchen, cool bedrooms, and passages and living rooms that were heated on special occasions. Today, standardisation required of building regulations and higher living standards have entailed the requirement that all rooms in the house must have the same temperature around the clock and all year round. This is one example of how standards for indoor climate and heat loss can contribute to maintaining a homogenous indoor climate that blurs man's experience of the outdoor climate, and undermines efforts to reduce fossil fuel energy consumption.

The division into different temperature zones can be a resource-saving layout, in which parts of the house are solely heated by passive solar heat, and where it is possible to shut off for use and heating of parts of the otherwise heated living areas. In an experimental building project from 1985 by Aude, Lundgaard, Sørensen and Rotne, the architects reduced energy consumption by utilising the Sun's heat radiation and applying zone division. The buildings are divided into three zones that can each function independently of each other by means of internal insulating shutters.

Fig. 6.17 a-b
THERMAL STABILITY
At the Öijared Golf Club, soil on the roof and granite
in floor and walls contribute to the thermal stability.
Constructed in 1986-88. Architect: Gert Wingårdh.

Fig. 6.18 a-b
MONT CENIS
In Herne, Germany, a mini town is covered by a
12,600 m² glass roof with solar cells in a cloud
formation. Zone division on a large scale.
Architect: Jourda and Perraudin, 1999.

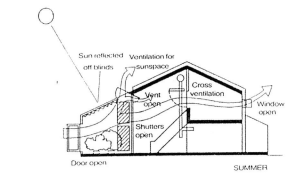

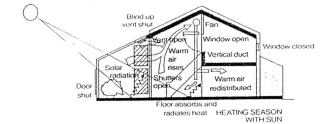

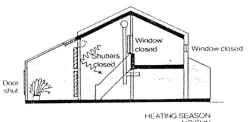

A highly insulated zone faces north; in the centre there is a common heated living zone, which receives a temporary heating contribution from a south-facing unheated greenhouse, which is ventilated and can be separated from the rest of the house by means of a glass section and sun curtains. Thus, the house can be fitted out and used in step with the changing outdoor climate.

One example of a large-scale zone-divided house is found in Herne, Germany. This is where one of the world's largest photovoltaic cell controlled buildings is found: The Academy in Mont Cenis from 1999. A gigantic glass box covers an entire small community with a roof area of 12,600 m², of which 10,000 m² are covered by photovoltaic cells, placed with different density as permanent clouds on the town's sky. The photovoltaic cells ensure an annual production of approx. 1,000 MWh or 130% of what the building complex itself needs. Hence, it is able to export 30% of energy production to the electricity grid. The glass covering creates a 'microclimate' similar to the Mediterranean climate. The buildings beneath the glass box can be constructed quite simply, almost like indoor spaces, protected against wind and rain and with a higher 'outdoor temperature'. In an energy perspective, the large passively heated indoor areas achieve energy savings of 23% over normal construction. During winter, the buffer zone takes the edge off the indoor climate, during summer the photovoltaic cells together with sun shading and deciduous trees along the east façade provide shade. Ventilation takes place mechanically driven by a central meteorological station. Roof and façade elements can be opened and ensure sufficient air flow by natural means. Further fresh air is supplied directly via underground ducts using geothermal sources – during summer as cooler air and during winter as warmer air.

Insulation

Just as heavy stone materials with great heat accumulating capacity may inhibit or delay temperature changes, very light, air filled materials may reduce heat transfer through an external wall. Trapped air is a very bad heat conductor, and insulation materials are materials that can surround air by means of very little matter. In cold climates, insulation will be intended to maintain heat. In warm climate zones thermal insulation is used to prevent overheating. Generally speaking the same insulation material can be used for both purposes.

Heat storage

Good utilisation of the energy potential of the mass presupposes knowledge of the matter's heat capacity or the matter's ability to absorb and emit heat. The heat capacity of a material increases with its density. One exception from this rule is water, which in relation to its density has very large heat capacity. This means that it takes large amounts of heat to increase the water's temperature. On the other hand, this is also what makes it possible to store large amounts of heat in water. Hence, water is a useful material to use to stabilise indoor temperature as long as

Fig. 6.19
ZONE-DIVIDED DWELLING
Divided into three thermal zones the houses in Greve, Denmark, can be used flexibly throughout the year.
Architect: Aude, Lundgaard, Sørensen and Rotne, 1985.

Fig. 6.20
INSULATION
Rockwool's headquarters in Hedehusene, Denmark, feature heavy insulation of the west-facing external wall with a rain screen of toughened glass.
Architect: Vandkunsten, 2000.

Fig. 6.21
THE ROCKY ISLAND OF CHRISTIANSØ BY BORNHOLM
The summer heat is stored in the rock, prolonging the summer season with a warm autumn.

evaporation does not take place. Solar heated water is often used for this purpose.

On the scale of landscape, the Danish island of Bornholm is an example of mass that can store energy and release it again. Because of its mass, a rocky island with great heat capacity and density is able to store the summer's heat and release it again in step with dropping temperatures. The radiation from the island prolongs the warm summer.

In buildings, the heat can be stored in heavy structures, or special storage areas made of stone, sand or water can be incorporated into floor, cellar, roof or wall structures.

The greenhouse – glass and mass
Sun terraces, glass houses, atriums and conservatories are all efficient solar collectors. The greenhouse in Bicton Park, Devon, is an example of utilisation of the greenhouse effect of glass. Soil, plants and a massive elongated south-facing wall on the northern edge combine to collect the rays of the Sun. The solidity of the wall helps to even out the temperature differences throughout the day and is used to absorb and store solar heat in the course of the day. Heat is released to the internal space during the night. Long wave heat rays are retained by the glass, raising the air temperature. This allows growing tropical plants in an otherwise hostile climate.

Uninsulated sun-facing wall – Trombe wall
The Trombe wall is named after the Frenchman Félix Trombe. Quite simply, it consists of a high-density heat-accumulating wall painted black. The wall is placed behind a sun-facing glass surface separated by an air space. The air in the gap serves as insulation, but the gap can also be used as a ventilation space. In the latter case, air circulates by hot air rising and being encouraged to flow into the house via an aperture at the top and out again via an aperture at the bottom of the wall. The heat-accumulating wall stores the day's solar heat and releases it to the rooms behind during the night. The time offset depends on the density of the wall.

Insulated Trombe wall
The Trombe wall depends on the Sun and the outdoor temperature. At a college in Windberg near Munich, architect Thomas Herzog has developed the principle further. Here, the south-facing wall behind the glass is constructed by concrete blocks painted black and covered with translucent insulation. Mechanically controlled roller shutters suspended in front of the glass surface are activated by overheating and rolled down during the night to prevent heat from disappearing.

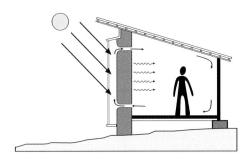

Fig. 6.22
TROMBE WALL

Fig. 6.23
STORAGE IN WATER
The dome over the Reichstag building in Berlin stands as an icon to its renovation completed in 1999. The building is both heated and cooled by geothermal storage at different depths underground.
Architect: Foster + Partners.

Fig. 6.24
HEAT STORAGE
In the greenhouse at Bicton Park, Devon, daytime solar heat is accumulated in the heavy back wall of the greenhouse, from which it is released again during the night.

Fig. 6.25 a-b
THERMAL EQUALISATION – THERMOLOGICA
Ship building iron plates as façade cladding sunk into the
ground ensure a warmer façade during winter and a cooler
façade during summer. Dwellings, Egebjerggård, Ballerup,
Denmark, 1996.
Architect: Dissing + Weitling.

Wall with phase-changing material

The Danish experimental project 'Boase' (*living oasis*) has used *climater,* a phase-change material produced in Sweden, as the basis for developing a shutter that can store the heat of the Sun. It is designed as an integrated architectural element in the Boase house and can be placed in front of the window, where it can store the solar heat of the day. *Climater* is a fabric that changes between fluid and solid states depending on the temperature in the room. When the temperature exceeds 24°C, the material gradually turns fluid. This process accumulates heat. Conversely: When the temperature drops to below 24°C, the material turns solid again and releases heat into the room.

RECENT DEVELOPMENTS

The idea of the lightness of industrialised construction which has been encouraged by internationalism has led to a minimisation of building structures. As a result the potential of heavy materials as a heat-accumulating and climate controlling factor in new buildings has been increasingly ignored, especially in the new economies of Asia as well as in the Middle East and the USA. However, many of the latest developments coming out of international architectural offices points towards a more complex understanding. There is a new interest in combining light and heavy building materials with buildings which employ large glass sections with heavy heat-accumulating materials in recognition of the optional saving of resources. Engineers now acknowledge that buildings for both the world's cold and hot countries can benefit greatly from integrating the qualities of several millennia's experience of utilising the Earth's thermo stability and building materials' heat- or cool-accumulating properties.

Fig. 6.26
PHASE CHANGE
The Danish experimental project 'Boase' which probably will be built in 'Christiania' in Copenhagen, evens out high and low temperatures by means of elements with a liquid that goes from solid to fluid state around 24 °C. The phase change accumulates heat at rising temperatures and releases it again at falling temperatures.

Humidity and Precipitation

Peter Sørensen

The Solar System's living planet, the Earth, is called 'the blue planet' because of its unique atmosphere, which is related to the presence of water. Water is essential for all life and forms a vital part of human life in many ways. We protect ourselves against it and use it, enjoy it, look at it, reflect ourselves in it, listen to it, smell it, use it in housekeeping, for farming the land and heating our houses. Water is a fundamental aspect to consider in architecture, where houses have to be protected against its erosive and destructive forces. On the other hand, the fluid and reflecting character of water can be used as an inspiration for the design, construction and use of architecture.

Pure water is essential for plants, animals and human beings. Fresh water only makes up a very small part (approx. 3%) of all water on Earth. The majority is tied up in ice and glaciers, a small part as atmospheric water vapour in clouds, whilst the rest is found as ground water and in fresh water lakes. Clean water is an important resource, but water in the form of precipitation, rain and snow is also a climatic condition for settlement and construction. The form, intensity, amount and temporal interval of precipitation depend on the general global climate systems, location within the Earth's different climate belts and particular regional geographic circumstances.

In tropical humid areas with monsoon rain, the roof is the most essential architectural element, which protects against the heavy downpour like an umbrella. When positioning and designing a building, consideration is also given to the large amounts of water that will need to be drained off the façade. In dry areas, where rain only falls infrequently, precipitation is a particularly important resource for humans and animals. Here, buildings are designed to make it possible to carefully collect the limited amount of water and direct it to tanks via roofs and paved courtyards etc. and stored for later use. In Arabian culture in particular, water is synonymous with fertility, as symbolised by the oasis around the desert well. Water is also a symbol of purification and treated with great care as an element in classical Arabian architecture's perception and striving for a recreation of the paradisaical garden. Water is a resource, but also a climate controlling and aesthetic element that affects architecture at all levels – from the design of taps to urban and landscape planning.

The following illustrates three relations between water and architecture. The example from Alhambra shows architecture *shaping the water*. Due to its architectural appearance, the traditional building from Northern Thailand may be said to be *shaped by water*, whilst the house from Australia is *shaped for the water*.

Fig. 7.1
WATER AS AN ARCHITECTURAL ELEMENT
A murmuring waterfall can be a refreshing experience. Tadao Ando often incorporates water as a natural, central element in his architecture; here from the Garden of Fine Art, Kyoto, Japan, 1994.

Fig. 7.2
ARCHITECTURE SHAPING THE WATER
Paradisaical garden with water as the central aesthetic and climate controlling motif. The summer residence Generalife, Alhambra, 14th century.

Fig. 7.3
ARCHITECTURE SHAPED BY WATER
Traditional building in Northern Thailand adapted to a rainy and humid climate.

Fig. 7.4
ARCHITECTURE SHAPED FOR WATER
The house collects the water, which will be used for sprinkling during frequently occurring forest fires. Simpson-Lee House, NSW, Australia, 1994. Architect: Glenn Murcutt.

Fig. 7.5
WATER SURFACE
Nature's own reflecting and changeable material in integrated interplay with glass, stainless steel, travertine and polished granite. A classic architectural effect. The Barcelona Pavilion.
Architect: Mies van der Rohe, 1929 (re-erected 1986).

Fig. 7.6
WATER CURTAIN
In urban planning, water is an important environmental, recreational and architectural element. The flowing, splashing, cooling and reflecting character of water can be utilised in many different ways. The World Exposition in Lisbon, 1998.

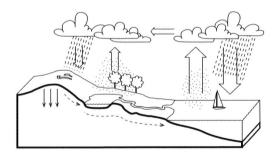

Fig. 7.7
THE WATER CYCLE
When water evaporates from sea and land, it is
cleansed and recirculated as precipitation that falls on
nature in an endless cycle.

Fig. 7.8
VISUALISED HUMIDITY
The air's humidity content varies according to the
temperature. Here, moisture is deposited as frost on
trees and bushes on a winter morning.

HYDRODYNAMICS

Water forms part of an endless cycle in symbiotic relationship with the Earth's rotation and an ever changing climate – experienced locally in nature as real weather or virtually on TV as a visualised global climate in fantastic satellite images of the Earth's atmospheric patterns. From atmospheric space, cloud formations and front systems in constant change and movement across sea and land areas are readily observed and communicated using today's advanced technology.

The word 'hydrodynamics' is composed of the two Greek words, *hydor*, meaning water, and *dynamis*, meaning force or display of energy, and it is the designation for 'the science of liquids in motion'.

The force behind the water cycle is the Sun. Not only does it deliver energy for water's evaporation, it also maintains the air currents that keep the cycle in motion. Evaporation takes place from oceans, lakes, watercourses, land areas and vegetation. When water vapours rise and are cooled, clouds are formed. Here, the small water particles densify to drops or ice crystals, which grow larger and heavier and fall to the ground as rain or snow. The precipitation is absorbed by soil and plants, diverted to watercourses or seeps into Earth's cavities and groundwater system. The upper part slowly flows towards watercourses, lakes and the sea from which it evaporates again, and the cycle is complete.

Urban planning and architecture, in their small way, contributes towards the water cycle, and water should be handled with care. When it is not understood, or when architects or engineers fail to design using natural systems, there may be flooding. As climate changes, so too does the pattern and intensity of rainfall. Water is a precious resource and hydrodynamics need to be as well understood as the issue of energy in buildings.

Water covers more than two thirds of the Earth's surface, and the oceans help ensure a certain thermal stability or inertia in the Earth's heating and cooling. The majority of global evaporation takes place from the surface of the oceans, and this water evaporation is also significant because of its cooling effect. Were it not for this cooling by evaporation, the Sun and the greenhouse effect would cause rapidly rising surface temperatures and a warmer planet.

In the temperate, coastal climate of Northern Europe, rain and fog formations with a high level of relative humidity close to 100% are common. Particular phenomena here include sea fog, which occurs by the coast at the encounter between sea and land, and local fog formations as ground mist, which develops in low-lying meadows during summer nights, when the cool of the night arrives, and the humid air is cooled and condenses. Dew forms on grass, fields, buildings and cars as a result of the humid air's encounter with cold surfaces. On cold winter mornings, moisture is deposited on trees and bushes, which may be completely white due to frost. When the air is heated, it is ready to absorb and contain more moisture again. Thus, the air's relative humidity content varies according to the temperature:

it is highest during winter and may vary from 90% to 40% in the course of a day, solely because of changing air temperatures.

WATER AND THERMODYNAMICS

Like other liquids, water has the ability to change from one state (phase) to another. This may be from ice, which is the solid state of water, to liquid form (water), to air of gaseous form (water vapour). In nature, the phase change or change of state from ice to water and from water to vapour happens through the supply or consumption of heat. The stored energy is released and emitted again as heat at a phase change the other way round, from vapour, which has a high energy level, to water and back to ice, which has the lowest energy level. This is a phenomenon that is particularly used in vineyards, which are sprinkled with water at the risk of frost during the important growing season.

Water is also used for building cooling by utilising water's evaporation energy and for air cleaning and binding of air particles through water humidification in air conditioning systems.

Water has a high heat storage capacity in comparison to most other materials. Water's ability to store heat is approximately twice that of concrete and granite (per volume). The ability of water to store and emit heat is used in heating systems with circulating hot water which emit heat by means of radiators (radiation heat) and convectors (air heat). An equivalent principle can be used for cooling by means of circulation of cold water.

Water's thermodynamic properties at the supply and emission of energy/heat at phase change from one form to another is interesting, but in terms of development relatively unheeded in construction. *Phase Changing Materials* (PCM) utilise this principle and hold new possibilities in connection with the development of façades and building energy optimisation.

The igloo is an example of exploitation of water in its solid state, tectonically constructed using porous, insulating blocks of snow set into a cohesive optimum form. It is an example where one material resolves all significant structural and climatic tasks of the building façade and roof. Snow and ice are not used for lasting architecture, but architectural and artistic experiments are known from the ice sculptures and buildings that are made and can be experienced during winter in the polar areas – the ice hotel in Jukkasjärvi being one of the best known examples.

HYDRO COMFORT

The physical survival of man and animals in nature is connected to the ability to register changes in the surroundings by means of the senses – sight, hearing, smell, touch and taste. Man can sense and react in relation to temperature changes, but is not equipped with a 'humidity sense' as such, whereby humans can register the invisible humidity content of the air. The air's relative humidity

Fig. 7.9
IGLOO
The porous, insulating dome of the igloo is built by blocks, carved out of firm snow, which in one material constitutes climate control and structure.

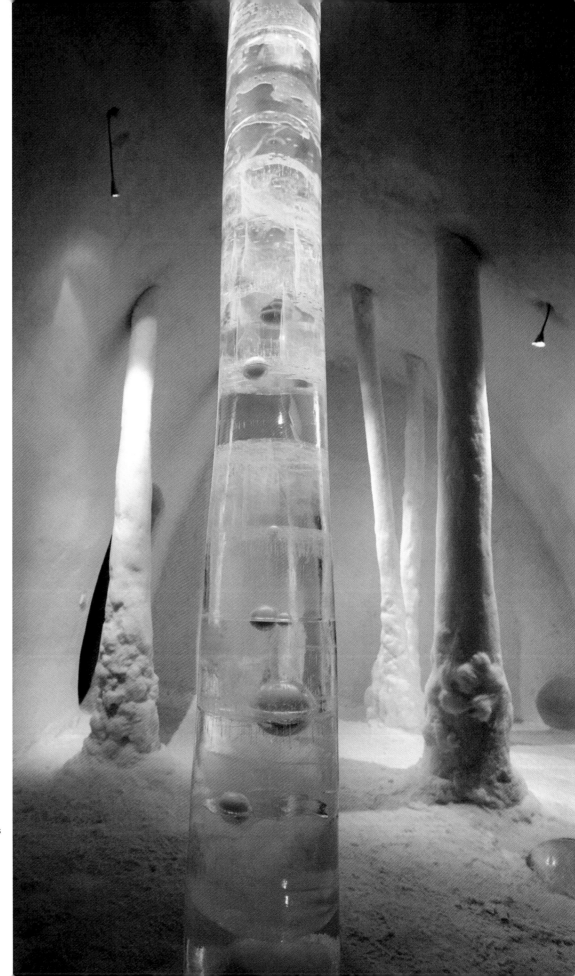

Fig. 7.10
SOLID WATER
The ice hotel in Jukkasjärvi is constructed
every autumn with the participation of a
large number of artists who design interiors
and windows in clear ice. Every spring, this
temporary architecture disappears.
Artist: AnnaSofia Mååg.

content can vary within a wide range (from approx. 20% to 70%) without any substantial physical significance or relevance to the experience of the air quality. However, high humidity buried in the fabric of buildings can cause problems such as timber rot due to trapped condensation.

High temperatures are experienced differently in dry desert areas such as Egypt and humid warm areas such as Panama in Central America. In Panama the high humidity prevents heat emission by means of sweating and surface evaporation from skin and lungs. Similarly, the experience of cold in a dry continental climate differs from the experience in a humid coastal climate where cooling happens faster. So, although man does not have a humidity sense as such, a high level of humidity increases the sense of discomfort in relation to both cold and heat.

One of the ways in which man's heat exchange takes place is through water evaporation from the body and the exhalation of air. In order to maintain a thermal balance in relation to the surroundings – regardless of activity level, cold or heat – moisture emission must be able to happen freely from the skin, through clothing and to the surroundings. In other words, the body needs to be able to breathe.

When dressing, it is therefore important that the innermost part of the clothes can transport moisture away from the body, and that there are no dense moisture-stopping layers outwards. Even though plastic is waterproof it is not suitable as rainwear, because it stops the body's moisture emission. The accumulation of moisture on the inside may result in uncomfortable cooling of the body, as water (moisture-saturated clothes) is highly thermally conductive and results in a greater direct contact surface between the skin and the cold air. This has led to the development of new synthetic fibre materials for outdoor wear, which is both rainproof and able to breathe, i.e. water permeable. The same considerations can be made concerning air humidity and the building's climate envelope, which is often built up layer by layer with cladding, absorbent, insulating, windproof and rainproof layers, corresponding to clothing.

The challenge today is to make the climate screens and façades of the future more flexible and able to adapt to the variability of the climate and the changes of the seasons and the day, just as we change our clothes.

In principle, the façade or climate screen only needs to be moisture-proof when it rains, insulate when it is cold, receive solar radiation when energy contribution is needed and shield against the same when there is a risk of glare or overheating.

Another perspective – if the clothing analogy is followed again – is that the façade's complicated structure of specific individual layers each with their optimised functions can be simplified. Technological developments suggest the emergence of new construction materials that perform several functions at a time – e.g. rain proofing, moisture filtration, insulation and transparency. New multifunctional membranes, development of nanotechnology etc. can lead to groundbreaking changes in the way we build. One parallel is seen in the continual technological development

Fig. 7.11
THE CONCEPT OF LAYERING
A concept of clothing and a concept of layering for an optimum control of moisture, temperature and wind between body and surroundings. The concept also applies to the layered structure of the climate screen and the façade.
Illustration from Craft brochure.

Fig. 7.12
MULTIFUNCTIONAL MEMBRANE
Translucent screen made of ETFE membranes, whose form is retained by slight differences in pressure as air filled cushions. Eden Project, Cornwall, 2001.
Architect: Nicholas Grimshaw & Partners.

Fig. 7.13

WATER VAPOUR DIAGRAM (MOLLIER DIAGRAM)

The diagram shows the connection between the air's water content, relative air humidity (curves) and temperature. The 100% curve is the air's dew point (the point at which water vapour condenses).

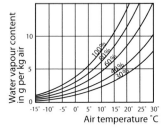

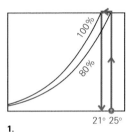 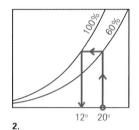

1.

In a room with a temperature of 25 °C with a high relative air humidity of 80% (e.g. a bathroom), air will condense on surfaces below 21 °C.

2.

In a room with a temperature of 20 °C with a relative air humidity of 60%, condensation may form on surfaces that are 12 °C or less.

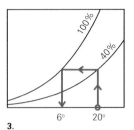 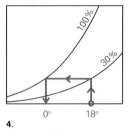

3.

If the air in the 20 °C room is replaced by outdoor air, relative air humidity is reduced to 40%, which at 20 °C will not condense on surfaces until they are 6 °C or less.

4.

In a room with a temperature of 18 °C and a low relative air humidity of 30% (e.g. a dry storage room), there is no risk of condensation until the temperature drops to below freezing.

of glass and glass façades, or the skin of aeroplanes, and the significance this has had in the last 30–40 years on architecture.

Putting aside the technology of water protection, a light shower of rain or a murmuring waterfall can be a refreshing sensuous experience on a warm summer's day. Architects and designers must be conscious of the climatic phenomena and influences we sense or experience outdoors in nature. These are similar to the laws that apply to a climatically conditioned, sensuous and aesthetic design of the spaces and details of architecture.

HUMIDITY AND INDOOR CLIMATE

The indoor climate's humidity content cannot immediately be registered; however, it has great significance both for the experience of the indoor climate quality and for health. Most often, large, light and spacious rooms feel comfortable, whilst smaller, low-ceilinged rooms with dense surfaces quickly feel stuffy. Climatically speaking, the large air volume has great natural capacity and therefore inertia towards changes in the air's humidity and temperature, whilst the small dense room is dependent on the regulation of air change and temperature.

Humidity is supplied to the indoor climate in many ways. At normal use, a dwelling for a family of four will supply humid air to the house corresponding to about 10 litres of water per day. High air humidity is a strain on the indoor climate. It is therefore a general prerequisite that dwellings are ventilated through air renewal 0.5 times per hour, so that humidity is reduced. Moisture deposited on the surfaces and materials of buildings may cause algae or fungi attacks, which may constitute a health risk. The risk of humid air condensing on windows and cold areas of external walls (thermal bridges) can be read in a water vapour diagram (Mollier diagram).

Problems and damage due to humidity may occur both on internal surfaces and hidden in the structures. Special consideration and use of vapour barriers is necessary if rooms and surfaces are exposed to great relative humidity (above 70%).

In bathrooms and other particularly moisture-exposed rooms, the humid air needs to be removed by venting, and dense, moisture-resistant materials and surfaces must be used. Furthermore, special membranes must be built into moisture-sensitive structures, and drying of the room and an adequate surface temperature must be made possible by means of heat supply.

HUMIDITY AND CONSTRUCTION MATERIALS

Materials such as metal and glass are completely solid and unable to absorb humidity. Even so, humidity-proof materials such as steel that is not protected against corrosion will rust when exposed to high air humidity. Other materials that are porous can accumulate humidity from the air and thus have so-called hygroscopic properties.

Hygroscopic properties in materials used for façades may be both an advantage and a disadvantage. Only very few materials are resistant to permanent high humidity strain. Many types of stone and porous bricks in themselves tolerate water-saturation, but in combination with frost, the risk of erosion and deformation increases. By burning a porous ceramic or baked brick at high temperatures, it is possible to change the pore structure. Through sintering, the material's pores are closed and the brick becomes frost-proof.

Wood is an example of a hygroscopic material that is capable of absorbing considerable amounts of moisture, as are many other organic materials. In changing moisture conditions it is necessary to consider that wood works and that the moisture content of wood should not exceed 20%. If the moisture strain exceeds this limit, wood and wood-based materials will be attacked by rot and fungi and are more prone to insect attack. Through careful choice of materials, correct climatic detailing and surface treatment to protect against soaking and ensure drying (constructive wood protection), the durability of wood façades is good despite the hygroscopic properties.

Some materials are capillary i.e. they are able to absorb and transport moisture – almost like a wick or a piece of blotting paper. Capillary transport of water takes place through capillary effect in porous materials from areas with high moisture content to less moist areas, from which the moisture can evaporate. The façade's encounter with the ground is particularly exposed. Water and moisture absorption is prevented by choosing moisture and frost-resistant materials and cladding, which will divert water off the building, and any perimeter drain. Rising soil dampness is countered by use of damp-stopping membranes, damp proof courses and capillary-breaking layers (gravel or expanded sintered clay).

Very hygroscopic materials such as wool or other comminuted organic fibres absorb moisture quickly and distribute the moisture into the material's large porous surface, from which it can easily evaporate again. The ability of materials to dry quickly is related to porosity and thickness. Thin organic cladding materials dry quickly through water evaporation, both from the exposed external surface and from the ventilated inside.

HUMIDITY AND FAÇADES
The weatherproofness of façades is about the interplay between construction, building design and external climatic influences – especially precipitation, humidity and temperature variations – and their ability, individually or in interplay, to destroy or impair the façade's durability.

The concept of *weathering* is used to describe this climate-conditioned patination and destruction process, which the building passes through over time. However, the concept is also a more active description of the idea of designing the particular building parts and details of the roof, overhang, coping, sills, drip edges, foundation walls etc., which are important in terms of controlling patination and preventing uncontrolled, rapid deterioration of the façade. This means that *weathering* can also

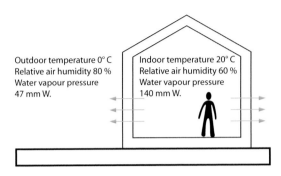

Fig. 7.14
ROOM DAMPNESS AND CONDENSATION
Water vapour pressure from the inside out to the façade. The connection between temperature and relative humidity can be seen in the diagram below.

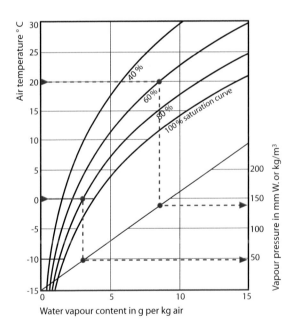

Fig. 7.15
WATER VAPOUR PRESSURE
A water vapour diagram (Mollier diagram) describes the relation between the air temperature (°C), the water vapour content (g per kg air) and the corresponding vapour pressure. Once the air temperature and the air humidity are known, the dew point temperature can be predicted and read in the diagram.

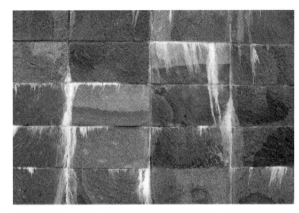

Fig. 7.16 a-b
MATERIAL AND HUMIDITY ABSORPTION
a. Stone is humidity-resistant but patinates especially in the external wall, as salts in the stone and mortar are leached out to the surface.
Beyeler Foundation Museum, 1997.
Architect: Renzo Piano.

b. The moisture-sensitive nature of wood is considered in the detail, in where the structure and the building are lifted off the foundation.
Timber engineering school in Biel, Switzerland, 1999.
Architect: Marcel Meili, Markus Peter and Zeno Vogel.

Fig. 7.17
THIN CLADDING MATERIALS
Residential complex in Dornbirn, Austria, 1996.
Architect: Baumschlager & Eberle.

be an expression of a conscious architectural handling of façade details in relation to the influences of weather.

Not all languages have an equivalent term to 'weathering' that comprises the expression of the important architectural discipline of combining technical insight, knowledge of materials and experience with detailed architectural design. The relationship between totality and detail, that is of such significance to the façade, apply to the building as a whole and is the key to its durability over time.

Where only a slow climate-conditioned destruction takes place (which does not require unusual maintenance or repair), the appropriate concept is 'patination'. When choosing façade materials and façade detailing, it is important to consider how the façade patinates. This means determining how a façade with the chosen materials and the technical execution can be expected to discolour or age and thus change in appearance in the course of time as a result of the climatic influences.

During extensive periods of time with high levels of humidity, materials and surfaces never quite dry out. In combination with frost and frequent temperature changes around freezing point (freezing point passages), moisture and precipitation constitute the most destructive climatic factor to which buildings are exposed. This is a typical condition for construction in Northern Europe's climatic zones, because of the coastal location with a very changeable climate. Buildings which use organic materials are particularly exposed, as lasting high humidity influence provides good conditions for the formation of algae, mould growth, rot and fungi.

Humidity and condensation may damage façades and building parts by:
– Rot and fungi growth in wood and other organic materials
– Destruction of materials and surfaces
– Deformation and frost erosion
– Water seeping in, reduced insulation capacity and water damage
– Deterioration of the load-bearing structure.

Facades are primarily affected by precipitation and humidity from outside, but warm and humid room air may also diffuse or flow into outer walls and roof structures. Humidity convection through flaws typically take place at the transition between wall and ceiling. If the incoming air encounters with structural parts colder than the dew point of the air, the air humidity deposits itself as condensation and in winter conditions as ice formation. In this way, great amounts of moisture are added, which can destroy the structure, reduce insulation capacity and result in rot and fungi.

The moisture diffusion can take place when there is a difference in vapour pressure between two air volumes (e.g. between inside and outside) that are brought into contact with each other through the pore system of a material. Moisture diffusion happens from the side of the structure that has the highest vapour pressure to the side with the lowest vapour pressure. Condensation may also occur in

Fig. 7.18
NORWEGIAN STAVE CHURCH
Norwegian stave churches are unique architecture and examples of how wood can be given long durability when it is chosen, used, detailed and maintained correctly in relation to the climate. Heddal stave church, constructed 1147–1242.

Fig. 7.19
PATINATION
Siedlung Halen, Bern, 1961. Architect: Atelier 5.

Fig. 7.20
ROOF DRAINAGE JAPANESE STYLE
Drainage of roof water is carefully detailed and executed in natural materials. Limited durability, but simple to replace. Classic Japanese detail.

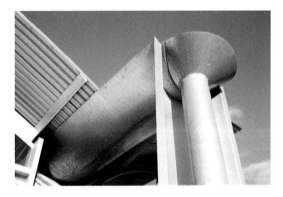

Fig. 7.21
ROOF DRAINAGE AUSTRALIAN STYLE
Here, rainwater is an important resource, which has defined the shape of the roof and the exposed downpipes. Magney House, Bingie Bingie, NSW, Australia, 1984.
Architect: Glenn Murcutt.

composite structures that contain diffusion proof layers in the wall's cold part or external surface.

Humidity and condensation problems in the façade structure are countered by considering choice of materials, façade design and details in relation to a series of simple principles.

Design principles in relation to humidity:
– Layer-divided façades to be constructed without cold bridges – as layers of clothing
– Internal surfaces and joints must be airtight
– Greatest vapour pressure resistant on the inside – decreasing vapour pressure resistance outwards
– Ventilated cavity between insulation and external cladding
– Rain shield with proper water drainage
– External joints to be completed as two-step joints with pressure-equalized cavity
– Water-draining flashing around façade apertures
– Condensation-insulated perforations of climate screen.

The book *On Weathering – The Life of Building in Time* written by Mohsen Mostafavi and David Leatherbarrow deals with the complex nature of the architectural project with a special focus on its temporality, linking technical problems of maintenance and decay. Most decay and weathering is influenced and enforced by water in combination with chemical and hydro-thermal processes – in short: dirt and staining. On this, Mostafavi and Leatherbarrow writes:

Dirt and staining: can they by anticipated? Certainly they are inevitable, but can they be projected, or envisaged as a likely future occurrence; still further, can they be incorporated into a design project? Staining is often the result of the juxtaposition of two materials, stone and metal for example, as in many nineteenth-century industrial buildings. When copper oxidizes and is washed by rain, a green stain is formed on the surface of the stone directly below. Stains seep into the porous stone, altering and deforming the original surface with these permanent markings. This may seem to by a deviation from the original intention for the surface color and texture, and it may be construed to have resulted from a fault in the design; but to the degree that stains show a new encounter between previously unrelated materials in one building sited in a particular place, they might also allow for a discussion of its harmony.

Fig. 7.22 a-b
THE ROOF AS A HORIZONTAL 'FAÇADE'
The roof of the Beyeler Foundation Museum in Switzerland is the house's primary 'façade', functioning both as rain screen and as light source with a number of light and climate controlling functions. Architect: Renzo Piano, 1997.

Fig. 7.23
A MODERN FAÇADE
A plane and dense glass rain screen suspended in front of the building's real façade. The glass screen is very dense and durable, but without patination quality. Materiality has been added to the transparent glass by means of a printed pattern of green spots.
Rossetti, Basel, 1999.
Architect: Herzog & de Meuron.

Fig. 7.24
A CLASSICAL FAÇADE
A richly profiled façade, where cornices and profiles frame and protect the building's apertures, repel water and moisture and ensure the durability and beautiful patination of the façade.

Fig. 7.25
CLARIFIED FAÇADE DETAIL
Prefabricated sill integrated into a concrete block
façade. Few and simple materials and clarified
geometry is the way to long lasting solutions. Tribune
Review Building, Pennsylvania, 1961.
Architect: Louis Kahn.

Fig. 7.26
SECURE WATER DRAINAGE
The primary purpose of the window sill from
Brussels is self-explanatory. Water-draining
flashing around façade apertures is important.

Wind and Ventilation

Peter Sørensen

The rise in global temperatures and resulting increase in levels of tropical storms has shown that there is a connection between the steadily growing man-made CO_2 emission and changes in the Earth climate.

The heat radiation of the Sun, the Earth's rotation and the relation between sea and land areas are the decisive driving forces behind prevailing wind systems on each continent. Heating of the Earth is greatest around the Equator and less towards the poles. The great differences in global heating in combination with Earth's rotation create a system of convection currents in the atmosphere, which constantly seek to counterbalance temperature and pressure differences.

The global atmospheric systems and prevailing wind directions result in great regional differences in terms of wind load, wind direction, intensity and temperature. Geographical location, proximity to the sea or height differences bring about particular local or seasonal wind conditions, which can be read in the positioning and design of towns, just as local architecture may be designed on the basis of specific wind conditions. It may be vernacular buildings' orientation and roof shape that protect against the weather or provide maximum wind exposure in order to achieve the cooling effect of the wind ensuring a comfortable indoor climate.

Local building culture holds important empirical knowledge of how architecture – by means of passive systems and limited resource use – can adapt to varied and sometimes extreme climatic conditions. These evolutionary passive climatic systems and design strategies can be developed further and combined with new technology for a modern-day resource-responsible architecture.

Today, many building façades are dominated by flat plane closed glass and metal surfaces, which presuppose artificially created indoor climates, based on extensive use of energy-demanding ventilation and climate systems. From a present-day point of view, the systems are expensive to install and operate; they pollute through CO_2 emissions and are quickly outdated. In fact, mechanical ventilation and air-conditioning is one of the fastest sources of obsolescence in modern buildings. Buildings that are based on natural systems of climate control must conceptually be planned and managed accordingly, and this gives the façade new tasks as an interactive provider of air, light and heat in the critical transition between inside and outside.

Fig. 8.1
AERODYNAMICS
The aerodynamic wing on the roof of the GSW Office Building in Berlin is the building's symbol of natural ventilation. Using negative pressure, the wing creates increased air renewal in the vertical continuous double façade through which the offices are ventilated hereby saving on energy use. The characteristic coloured shutters control light, shade and incident sunlight.
Architect: Sauerbruch Hutton Architects, 1999.

100 % 51 % 49 % 10 % 3 %

Fig. 8.2
WIND RESISTANCE PROFILES
A plane surface positioned perpendicularly to the direction of flow gives maximum air resistance and creates negative pressure and turbulence on the back of the plane. A spherical form reduces resistance by half, and the drop shape reduces air resistance to under one tenth.

Fig. 8.3
OPTIMISED DYNAMICS
In the Tour de France individual time trial, achieving the optimum relationship between force, weight and air resistance is crucial. This image is of Lance Armstrong wearing an aerodynamic helmet and using a carbon fibre bicycle. Within a decade innovation like this spread into other industries, including the construction industry which in general is the most traditional and conservative of the worlds major industries.

AERODYNAMICS

The aerodynamic nature of the wind and the relation between wind and building design has primarily been studied in the context of structural considerations. Aerodynamics have been explored particularly within the aviation, wind turbine and car automobile industry, but many of the aerodynamic principles and solutions that have been developed can also be applied in the world of architecture and with great potential in the field of sustainable design.

Aerodynamics is about airflow and forces that affects a body in a field of flow. Air resistance depends on an object's cross-sectional area, shape and surface friction. When the wind moves across a shape or a building, a negative pressure or force occurs, which increases exponentially with the wind speed. When the wind speed around an aeroplane wing profile increases, the air is forced to flow quicker over the profile than under the profile. This creates the pressure difference that is the force that gives the aeroplane its lift. These principles are used in the ventilation of buildings where the channelling of air currents is used to increase the effectiveness of natural ventilation, especially in office and school buildings.

Wind turbines, which are constructed according to the same aerodynamic principles as aeroplanes, are developed to produce environmentally sound electrical energy from wind. Recent design and technological optimisation has improved the performance of wind turbines to the point where 80% of the primary wind energy can be converted to electricity. Car models are tested in wind tunnels to reduce air resistance and thus reduce energy consumption, but also to develop new models without wind roar with an aerodynamic design that constitutes a competitive sales parameter that adds value to the product.

Within many sports, manufacturers and designers use aerodynamic testing to develop sports equipment that is aerodynamically and functionally optimised for different sports. Careful analyses aimed at reducing weight and air resistance lead to completely new materials and techniques. Air friction and weight are reduced as much as possible in for instance the development of bicycles and figure-hugging, breathable membranes or windbreakers, which may be the decisive difference when chasing the marginal seconds in many competitive sports. Professional athletes, joggers and cyclists alike demand the best and most advanced equipment.

From the world of architecture, there are examples of how wind tunnel tests and simulations, calculations and IT visualisations of wind data in and around buildings have led to new aerodynamic architectural design, for instance in the work of Future Systems and Foster + Partners. Research and development in new composite materials, membranes and lightweight structures emerge and find their way into building structures and new dynamic façade screens.

Fig. 8.4
SWISS RE IN LONDON
The aerodynamic shape of this high-rise office building was developed and tested by means of digitalised wind calculation models and tests that documented the characteristics of the shape's wind environmental properties. With less wind resistance and hence reduced turbulence there is significantly improved wind conditions around the building at pedestrian level, as well as reduced energy consumption within the building.
Architect: Foster + Partners, 2004.

Fig. 8.5
SWISS RE IN LONDON
Illustration: © BDSP

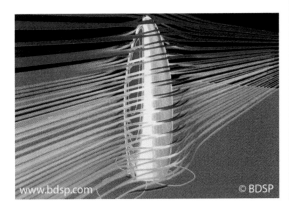

Fig. 8.6 a-b
THE CITY HALL IN MARSEILLES
The 'blue city hall' in Marseilles applies the aerodynamic principle that wind sheers off the building. The building's profile is designed according to the two seasonal winds, i.e. the strong, cold Mistral from the Alps (mainly northerly) and the warm and dusty Sirocco from Northern Africa (mainly southerly). Architect: Alsop and Störmer Architects, 1994.

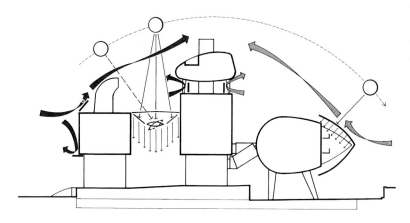

AEROCOMFORT

The ability of humans to spread across the globe was mainly by means of the protection afforded by clothing. Clothing protects the body against wind, drying out and cooling, and ensures bodily comfort. The building and particularly the climate screen may be considered an extension of clothing. Like clothing the building façade is to protect against climatic influences and ensure indoor comfort. However, like clothing, the building façade needs to be flexible so that different climatic conditions can be moderated. Ideally the climate screen, like an overcoat, should be changeable and able to adjust dynamically to meet changing external conditions.

In humid warm climate areas, a breath of air on a hot day is pure bodily delight. The hammock, which allows air movement immediately under the body, originates from the tropical, humid zones, where it is commonly used during warm nights. Physical comfort is entirely dependent on the air's movements and the wind's cooling effect directly on the body's surfaces. Cooling happens both by convection and by evaporation through perspiration.

In cold climatic areas exposed to wind, the issue for both man and buildings is to find shelter to avoid the wind's cooling effect, which rises exponentially with increasing wind speed. Hence it is important to ensure external wind-proofing and indoor air-tightness to avoid asymmetric cooling of the body due to draughts. Indoor air circulation is good for health, but it is important that it is controllable by users.

Regional traditions in architecture offer many examples of how basic comfort requirements are met in very different wind and climate conditions – with examples ranging from cold areas exposed to strong winds, where the issue is wind protection, to very warm areas, where optimum wind utilisation for building ventilation is desirable. Different vernacular patterns of building use different technique to determine the indoor climate and temperature. Hence traditional architecture varies accordingly to wind patterns and intensity, not just to temperature.

Sustainable and eco-friendly architecture seeks to explore and utilise local wind conditions, partly by dealing with the destructive and negative forces of the wind through wind protection, and partly by exploiting the wind's positive potentials through wind utilisation. The latter finds expression in two main areas – positive wind for ventilation especially of deep-planned buildings, and through local wind generation of electricity. Micro-scale wind power offers much potential in the future as society moves towards a post fossil fuel economy.

Wind protection

Anonymous architecture often finds its form shaped by the need for wind protection in the rural or urban context. At the coast, it is possible to experience how wind dynamics are created in a symbiotic exchange of air based on thermal differences between sea and land. Throughout the day, the Sun heats up the Earth, while the sea remains cool, and when the warm air rises, the cold breeze flows

Fig. 8.7
BODY-HUGGING WIND UTILISATION
A breath of air on a hot day can be sheer bliss for the body. In humid warm areas, the hammock allows for a body-hugging utilisation of the wind's cooling effect.

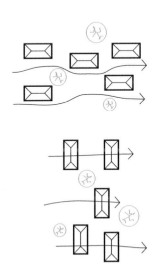

Fig. 8.8
WIND PROTECTION AND WIND UTILISATION
In cold areas exposed to wind, the issue is wind protection; in warm and humid areas, the issue is wind utilisation.

Fig. 8.9 a-b-c
WIND-ORIENTATED URBAN STRUCTURE
In the village of Sønderho on the island of
Fanø, Denmark, the design of the houses,
and the town plan with buildings lying east–
west, reflect the need for protection against
the prevailing wind. Compare with the wind
rose shown in fig. 8.10.

Fig. 8.10
WIND ROSE
The wind rose sums up wind direction and wind force
on the west coast of Denmark at Fanø measured over
a 10-year period. The main wind direction oscillates
between SW and NW, which also gives the highest
wind speeds. The wind pattern would be quite
different inland and on the east coast of Denmark.

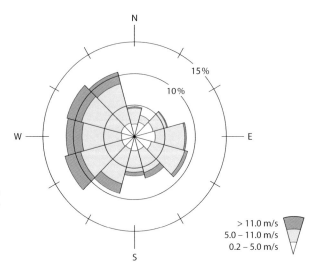

inwards over land to replace the warm air. During the night, the opposite happens, and the wind turns. These patterns make costal architecture subtly different to elsewhere and interesting to study by architects building near to the sea.

If a building is to be located in a coastal area exposed to strong winds, it is necessary to take the weather into consideration when designing the building and its immediate surroundings in order to reduce the force and cooling effect of the wind and to create shelter. One example of this is the characteristic windbreaks and large tall trees by farms in Western Jutland. Wind load and cooling can also be reduced by means of the type of building, building orientation and interior layout.

It is characteristic of the village of Sønderho on the island of Fanø, Denmark, that the rectangular houses are placed closely together and orientated east–west according to the prevailing west wind. The houses are low and of a simple, closed shape with their gables facing into the wind. Outside, banks and fencing shield small gardens. A porch or a small entrance hall form a wind lock that serves as a climatic buffer zone towards the south-facing, heated sitting room for winter use, while the room for summer use is lying to the north, where you have the early morning and late evening sun.

In cold climate areas, extensive, uncontrolled air renewal will feel unpleasant because of draughts, and will cause a significant heat loss and thus excessive energy consumption. In the wind-exposed and rainy coastal areas in much of Western Europe, a system of glass bays has been developed, whose primary purpose is to protect against rain, wind and cooling winds. The characteristic seaside façades in La Coruña consist of a transparent glass screen with glazing bars in front of heavy stone façades with traditional windows and doors for ventilation and access. This creates a climatic buffer zone through which it is possible to ventilate the rooms behind, and utilise the benefits of the sun and view. These façades may be seen as forerunner of the present day's more advanced naturally ventilated double glass façades, which utilise glass and thermal mass to achieve comfort.

Wind utilisation

In contrast to wind protection – being able to use the clean fresh air and the wind's dynamics is a positive parameter in architecture. Whilst sailing ships use the wind as nature's own propulsion agent, and wind turbines create pollution-free energy, architecture primarily uses the wind to ensure air renewal and cooling by means of natural ventilation. People, buildings and towns depend on being able to breathe fresh air – but the wind can also be ascribed conceptual and formative significance in the design of buildings and urban planning.

Due to rising density, increasing temperature and air pollution, towns and cities are increasingly dependent on wind to ensure air supply and remove excess heat and polluted air. Manhattan is very energy, heat and air pollution intensive, but because of its location by the sea and wide street grid, coupled with large green airy spaces such as Central Park, it is quite a pleasant place to live and work. Despite

Fig. 8.11
RURAL WIND PROTECTION
Detached houses and farms in the landscape of Northern Jutland, Denmark, shelter against the west wind using the landscape, trees and hedges.

Fig. 8.12
VILLA ROTUNDA
Villa Rotunda by Palladio is situated high up and conspicuously, and by means of its location and design, makes optimum use of the cooling breeze regardless of wind direction.

Fig. 8.13
WATER COOLING
At the entrance to a traditional house in Thailand, you are met by cool drinking water stored visibly in porous clay jars. Shade and air circulation ensure cooling by evaporation.

the great density of the city, New York has a tolerable climate thanks to the all-pervasive street grid which helps carry fresh air from the costal locations through to the urban core.

The same thing does not apply to Tokyo, a city referred to in terms of the special 'Heat Island Phenomenon' – that is the unwelcome build-up of external temperature as a bubble over the city. The following quote comes from the Venice Biennale 2006, the theme of which was 'Cities in a Changing World'.

Global warming issues are compounded in Tokyo, where average temperatures have risen more visibly than in any other places in Japan and the rest of the world. During summer, the city becomes 'a heat island' due to increased emissions, the amount of asphalt that covers the city's surface, and the large number of high-rise buildings on the waterfront, that act as a barrier to the cool breeze from Tokyo Bay. The 'cool island initiative', linking green areas in the city, the practice of watering asphalt to cool it down, called 'uchimizu', the adoption of 'cool business' activities, and many other public and private small- and large-scale initiatives are now being taken to address the problem.

Air in buildings can be cooled naturally in various ways. Air can be drawn in via cold cellar rooms or subterranean ducts or hollow building parts in walls and floors. In each example air releases its heat and is cooled by convection and conduction against the cold surfaces of the building. There is a long tradition of cooling by such means. In Palladio's Villa Rotunda, the connection between landscape and building design is clear and distinctive and driven by climatic considerations. The relationship to the place and the climate is emphasised by the way in which necessary ventilation has been achieved as an integrated part of the architectural solution using the central rotunda. The detached building is exposed on a hill top and is able to capture cooling breezes from any direction. The central circular space is completely separated, both geometrically and physically, from the building's exterior. The dome's lantern provides both light to the central space and serves as exhaust for the air that is supplied via the cool cellar rooms of the building and led through the rooms to the central space via solar and wind driven cross-ventilation.

Air can also be cooled and purified by water evaporation. Cooling by evaporation happens by heat from the air being used when water evaporates and turns from liquid state to water vapour. In India, one solution is to use frames with thick woven mats, which are moistened and hung in front of windows and doors, thereby cooling the ventilation air. The same principle is used for cooling of drinking water stored in porous ceramic jars. A reflective water surface in front of a building is both beautiful and pleasant, because the air that passes across the water surface is humidified, whereby it is cooled and purified of dust particles.

Fig. 8.14 a-b
COOLING BY AIR HUMIDIFICATION
A west-facing glass façade utilises passive solar energy but
at the same time creates overheating problems. These are
naturally resolved by using cooling by air humidification via
thermal lift in the three-storey room. Photovoltaic cell
energy controls the glass sections' movement and an
additional external solar screen.
Energy Research Centre, Gelsenkirchen, 1995.
Architect: Kiessler + Partner.

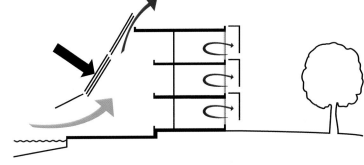

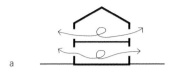

a

b.1

b.2

b.3

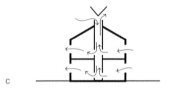

c

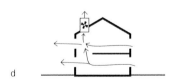

d

Fig. 8.15
PRINCIPLES FOR NATURAL VENTILATION

NATURAL VENTILATION

In temperate climate areas, the wind is used quite naturally for ventilation of houses and ordinary buildings. Ventilation serves three primary purposes:
– To ensure the supply of clean, fresh air – in an amount that is sufficient and which can be controlled.
– To ensure indoor climatic quality, comfort and well-being – excess heat, air humidity and polluted air are removed by means of adjustable ventilation.
– To cool the building and its stored thermal mass (e.g. through night cooling or air-conditioning).

On a spring day, pleasant internal conditions can be provided by simply opening a double door and allowing the fresh morning air into the room. But on a cold winter day, the door is not appropriate as the only ventilation possibility, as it is difficult to control air renewal, and thus the room temperature and heat requirement may suffer. In order for natural ventilation to work, it should be possible to air the room more evenly by means of smaller, adjustable apertures in the façade and via exhaust or neighbouring rooms out into the open. This means that natural ventilation presupposes a conscious spatial layout that takes into consideration how air is transferred from one room or floor to another and from one zone of the building to another. In dwellings, air transfer should always take place from less polluted rooms to more polluted rooms, i.e. from sitting room to kitchen or bathroom.

Night cooling is a principle that is used in connection with ventilation of office buildings. In daytime, excess heat from people and machines accumulates in the building fabric; during the night, the building is cooled down by complete airing. Exposed structure allows the night time cooling effect to be more readily exploited.

Natural ventilation can be planned in accordance with the following principles:
a. By means of *pressure differences across the building* – and through windows by utilisation of the natural pressure variations created by the wind.
b. By *thermal lift* – through chimneys and ducts (b.1), through glass façades (b.2) and through-going, vertical spaces and atriums (b.3) the 'chimney effect' is utilised as the underpressure that is created by the rising heated air. (Solar assisted ventilation is based on the thermal lift principle).
c. By means of *wind catchers and wind towers* – which work at both overpressure and underpressure.
d. By means of *hybrid ventilation* – where natural ventilation can be improved by being connected to mechanical systems. This is often called 'mixed-mode' ventilation (wind, solar, thermal and mechanical used together).

Fig. 8.16
MAXIMUM VENTILATION
The traditional house in Thailand is an example of a climate-adapted house built from bamboo and leave fibres, which are light and airy and ensure maximum ventilation. The lightness of construction results in a climate responsive dwelling.

Fig. 8.17
MINIMUM VENTILATION
The Nubian mud house adapts to the climate by having great mass and very small apertures, which are closed during the daytime when it is hot. At night, they are opened for ventilating and cooling of the building.

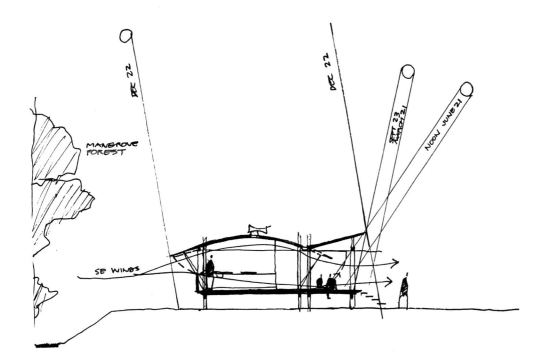

Fig. 8.18
CLIMATE SKETCHING
Sectional sketch for Marika-Alderton House, Australia, by Glenn Murcutt. This early sketch of the house section establishes the decisive climatic parameters for the architecture of the house. For such preliminary analyses, the cross section of a building is more valuable than the plan.

Fig. 8.19
MULTI-ADJUSTABLE FAÇADE
The traditional sliding Japanese window is incorporated into an adjustable façade offering a high level of flexibility of light and air in the transition between inside and outside. Such dynamic façades reflects respect and gratitude for nature, which is a key feature of Japanese architecture.

Fig. 8.20
MULTI-ADJUSTABLE WINDOW
The window is an integrated part of the room's layout with many possibilities for adjustment of light and air. Boyd Education Centre, NSW, Australia, 1999. Architect: Glenn Murcutt.

Ventilation by pressure differences

Natural ventilation across the building is achieved by making use of the pressure differences that surround the building, by placing air inlets in areas with positive pressure and outlets in areas with negative pressure. Normally, there will be positive pressure on the building's windward side and negative pressure on the leeward side and over flat roofs.

The pressure difference between the air intake and the position of the outlet determines the airflow dynamics, i.e. the speed and amount of air that is forced through the building. The speed will increase if the total amount of air is compressed in small sections, and drop when it passes larger sections or if the volume is increased. In exactly the same way as the current in a stream, which will run faster through narrow passages and slower when the stream widens, and it will almost stand still when passing through a lake.

Buildings in humid, warm climate areas need to provide maximum shade and ventilation to ensure through airing. Therefore, they are light and airy with large roofs; and normally surrounded by verandas and frequently raised above ground. Walls and floors are thinly constructed and often air-penetrable, and light and air are filtered by lattice shutters, which allow even the lightest breeze to pass. Light materials such as wood, bamboo and leaf fibres are used to avoid thermal heat storage. The air is directed into the building via shaded areas at body height above the floor – or through it with total cross-ventilation in order to cool the surfaces that are touched as much as possible.

Whereas maximum airing is important in humid, warm areas, it is necessary to create maximum separation between inside and outside in warm and dry areas. In such locations one finds heavy buildings made of stone, clay, mud or brick. Because it is warmer outside than inside, thermal cohesion by direct air connection between outside and inside could result in an uncomfortably hot building. The answer to this is to shut off window apertures and ventilation during the day when it is warm, and open windows and achieve maximum ventilation during the night. Airing by means of the cool air of the night chills the building's great thermal mass, which is then slowly heated again in the course of the day. This and the Japanese example demonstrate how the use of the buildings by the occupants needs to be in sympathy with the aims embedded in its construction. With industrialization and air cooling, using energy demanding air-conditioning systems, such traditions are fast disappearing in many parts of the world.

Windows and doors in traditional, closed façades adjust the climate between inside and outside – often with a special design that facilitates control of the building's air renewal. Experiments with new window types that can pre-heat air intake and which are equipped with a self-adjusting split valve have shown that the energy loss at ventilation can be reduced significantly.

The development of new window types with pre-heated air intake owes part of its inspiration to the windows and natural ventilation principles of the Winter

Fig. 8.21
SCANDINAVIAN WINDOW
The traditional Scandinavian window which primarily is oriented to the south for solar access, adjusts light, air and temperature in keeping with the changes over day and seasons.

Fig. 8.22
MEDITERRANEAN WINDOW
The Mediterranean window is the façade's multi-adjustable mechanism between the room and the surrounding world.

Fig. 8.23
WINDOW FROM CAIRO
The grating in the *mashrabiya* filters air and light and shields against direct sun and people looking in. The hotter the climate, the smaller the proportions of openings in the grate.

Fig. 8.24
DOUBLE GLASS FAÇADE
Double façades work by means of an extra layer of glass being added to the exterior façade. This facilitates new ways of handling heat loss, cold radiation, solar shielding and air intake/outlet as natural ventilation. Double glass façade in the office section.
Storage library, the Danish Royal Library, Amager, 1997.
Architect: Dissing+Weitling.

Fig. 8.25
CONCEPT SKETCH
Concept of the ventilated double façade in the storage section. For maximum stability in light, temperature and moisture, the books are stored in a dark heavy box of big thermal mass, separated by the buildings outer solar and rain protecting skin with a ventilated cavity in between. Storage library, the Danish Royal Library, Amager, 1997. Architect: Dissing+Weitling architecture.

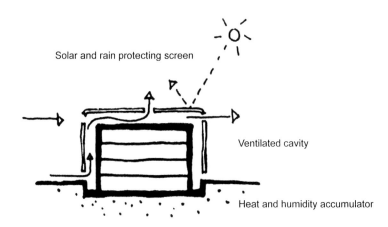

Solar and rain protecting screen

Ventilated cavity

Heat and humidity accumulator

Fig. 8.26 a-b
DOUBLE WINDOW
A new type of window that pre-heats ventilation air between two glass surfaces has been developed for residential buildings. Heat recovery, air filtration and sound attenuation are achieved by placing an aperture at the bottom of the outer window and at the top of the inner window. Jægersborg Water Tower, Denmark, 2006. Architect: Dorte Mandrup Arkitekter.

Fig. 8.27
VENTILATION CHIMNEYS
The steel ducts in the façade form part of an advanced energy-saving natural ventilation and cooling system. When solar generated thermal lift is used, cool air is drawn into the building from the north side or stored in hollow floor components. BRE Environment Building, Watford, UK, 1997. Architect: Feilden Clegg Bradley Studio.

Palace, St Petersburg, which utilise the gap between outer and inner glass layers to pre-heat the air needed for ventilation. Air is drawn in from outside through the double layers of the window at the bottom and let into the room at the top. Each room has a supplementary ventilation duct for air outlet integrated in the walls. The air intake is adjusted by means of a simple manual valve. The principle is based on the air being pre-heated when it meets the inner warmer glass surface where it then rises to ventilate the room through the top duct.

When the window is made of glass on the entire façade from floor to ceiling, a much greater direct contact is achieved between the room's interior and the surroundings. However, the glass façade also needs to fulfil all of the outer wall's traditional climatic functions, and as the window, it should be able to control a series of climate conditions between inside and outside. This applies not least to an increased need for ventilation because of incident sunlight and excess heat. Double façades work by means of an extra layer of glass being added to the exterior façade. By introducing a gap or a buffer zone between the inner and outer façades, it becomes possible to handle heat loss, cold radiation, solar screening, air intake and outlet in new ways through the façade as natural ventilation.

Ventilation by thermal lift

Natural ventilation can also be achieved by employing the principle of thermal lift, either in buildings' sectional spatial layout (using for example double-height rooms) or by using chimneys or air ducts. Thermal lift works by warm air rising because of its lesser density, whilst air that is cooled will drop correspondingly. This is a phenomenon that is often experienced from the draught near cold window surfaces.

'The chimney effect' is achieved by using thermal lift in a vertical duct, and the effect can be enhanced if the chimney's mass is warmer than the air. When the height of the rising air column is increased, the pressure difference increases and thus the air intake at the bottom also increases. If the outlet is placed high above the building's roof or ridge, the effect is further amplified because the wind that passes through the aperture contributes to creating a low pressure. Often solar space heating is employed to accelerate the process.

With thermal lift, negative pressure is created in the room, and this means that air renewal happens by means of fresh air from outside. The air renewal is simply controlled by window apertures or valves. Ventilation using thermal lift is particularly advantageous in double height rooms, central spaces or atriums, which are commonly found in office buildings and shopping malls. The same principle can be applied for double glass façades, which pre-heat and ventilate the rooms behind, often using a ventilation chimney at the top to enforce ventilation and avoid overheating in sunny seasons.

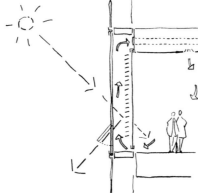

Fig. 8.28 a-b
DANISH BROADCASTING CORPORATION
Facade completed with a naturally ventilated double glass façade based on thermal lift. The outermost glass layer is suspended from the roof edge.
Photo and simplified diagram: Danish Broadcasting Corporation, Segment 2, 2006.
Architect: Dissing+Weitling architecture.

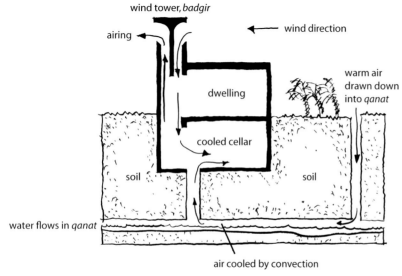

wind tower, *badgir*

airing

wind direction

warm air
drawn down
into *qanat*

dwelling

cooled cellar

soil

soil

water flows in *qanat*

air cooled by convection

Fig. 8.29
WIND TOWERS

The traditional wind tower, *badgir*, towers above
the city of Yazd in Iran and features apertures
which face in all directions in order to capture the
wind and direct the cooling air several floors down
into the building.
View over Yazd, Iran.

Fig. 8.30
SIMPLIFIED SECTION OF WIND TOWER

The house is ventilated by the wind tower, which draws up
cooled air from subterranean ducts (*qanat*).

Ventilation by means of wind catchers and wind towers

In the city of Hyderabad in Western Pakistan, the roof landscapes are characterised by the many angled panels protruding upwards as wind catchers. They are all pointed in the same direction with apertures towards the prevailing cooling '120-day wind' from the distant mountains to the north. The wind catchers are simply constructed of three timber panels and an adjustment rod. Through these, cooling wind is drawn downwards to each individual room. The shaft is equipped with a flap that can adjust the airflow and shut off completely during winter, as well as a metal grate that can keep out unwanted visitors such as birds. In contrast to modern-day cooling systems, this system is completely natural and independent of the local, (often failing) electricity power supply.

Another type of wind catcher, *malkaf*, is found in the Egyptian urban dwelling. It is a larger shaft that protrudes above the building and works as wind catcher with a large, wide aperture that collects the prevailing, relatively weak north wind and forces the air downwards through the building. The air is diverted by means of a lantern placed high up above the largest central room, enhancing the thermal lift of the airflow. The Egyptian wind catcher can be combined with an angled, grooved marble plate, known as a *salsabil*, with trickling water. It is designed so that the wavy pattern of the plate delays the water flow and furthers evaporation and hence cooling of the air.

Wind towers or *badgirs*, which are particularly common in Iran and Iraq, are a more advanced version of the *malkaf*. They work as both wind catcher and ventilation chimney and can be adjusted using flaps. The wind tower base measures about 3 x 3 m, and they are up to 7 m in height. The top part has tall apertures facing in all four directions. The apertures, which are separate along the entire height of the duct, can capture winds from different directions and guide the cooling air down into rooms up to two floors down. The cooling effect of a wind tower can be enhanced by water evaporation from moistened cloths placed at the bottom of the shaft aperture or by means of clay jars filled with water. The tower also serves as a ventilating chimney, drawing warm air up on the lee side because of negative pressure hence ventilation is possible even if there is no wind because of thermal lift aided by the heat of the sun. There may be several wind towers in one building and they are usually located adjacent to central courtyards.

In principle, this building type is organised with rooms and multi-purpose spaces, which can exploit variable climates. The occupants move around in the building depending on where it is comfortable to stay. The two floors with living quarters thus have seasonal utility, just as the flat roofs can be used as sleeping places during summer. The tradition of the building's organisation in relation to climate is understood by the occupants and hence management of environmental systems and culture are closely related.

The traditional house of courtyards and wind towers provides a form of architecture that offers spatial variation as well as functional flexibility. Its refinement in

Fig. 8.31
WIND CATCHERS
Wind catchers, Hyderabad, Western Pakistan.

109

Fig. 8.32
MODERN WIND CATCHERS
On the BedZed sustainable development building in Sutton, London (2002), wind towers ensure both air intake and air exhaust in one single fitting on the building's roof. They are adjustable according to wind direction. Architect: Bill Dunster.

terms of climate stands in contrast to much modern construction in the region. Most recent buildings are mechanically air conditioned, providing rooms that are proportioned and engineered for one specific purpose and with the same room temperature everywhere, regardless of orientation. A long tradition of building in sympathy with nature is thereby subtly being undermined.

Hybrid ventilation

If modern-day construction is considered in the light of sustainable development perspective, one of the main points in planning the building's service systems will be to look for the simplest solution rather than choosing complicated systems, which are expensive to purchase, operate and maintain. In small building projects, it is possible to apply knowledge of basic principles of natural ventilation from vernacular buildings to achieve both simple and sustainable ventilation. The low-tech solution does not require special mechanical installations or expensive climate control systems. Low-tech, sustainable solutions need to be incorporated early in the planning process since they are both easy and inexpensive to establish. Studies show that the human sense of well-being depends on retaining control of air, heat and light in the building. If you work in an office where it is not possible to open a window or adjust the level of light, you rightly feel a sense of discomfort and dissatisfaction with the building. This may affect your level of production or commitment to the company you are working for. However the use of natural ventilation in large buildings is complicated, and it is a prerequisite that the environmental principles are integrated into the architectural concept right from the beginning. This is of particular significance in the design of façades and the relationship between climate systems and the organizing and nature of interior spaces (depth, height).

Some of the challenges in the development of natural ventilation in modern architecture are exact control of air renewal and utilisation of the air's energy at heat recovery.

Formerly, the choice of ventilation principle for large building projects has been a question of either/or; either mechanical or natural ventilation. It is only natural that in large buildings where a multitude of human activities take place, there is a desire to establish a relatively stable climate. The only ventilation principle that has been able to ensure stable temperatures and air renewal is the mechanical principle, which has meant that in particular in relation to large construction projects, sustainable ventilation principles have been rejected. The natural ventilation principle has slipped into the background in favour of the mechanical principle.

Mechanical ventilation developed from being a simple technique that provides a room with a constant and uniform airflow to a more energy-optimised ventilation system with a relatively low energy load that recovers heat and controls air currents according to need. During the same period, natural ventilation developed from being a largely incalculable and uncontrollable possibility into a ventilation system that

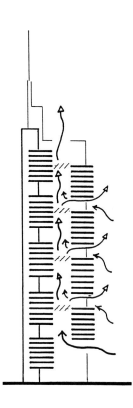

Fig. 8.33 a-b
HYBRID VENTILATION
In the 53-storey office tower, natural ventilation in central atrium
areas is combined with controlled air conditioning in office
areas. It is, however, possible to control the office areas
individually via the climate system and to replace air-conditioning
with natural ventilation by opening the windows. The central
atrium and the four-storey green gardens are offset up through
the building providing air by natural ventilation, arranged as a
spiral of interconnecting green spaces.
Commerzbank, Frankfurt, 1997.
Architect: Foster + Partners.

Fig. 8.34
VENTILATION TOWERS
The low energy project, which is based on a combination of natural ventilation and mechanical ventilation driven by solar cells, was developed in collaboration between Arup and Hopkins and supported by EU grants. The different spaces, some used for teaching, others for circulation, are arranged to take advantage of differential air pressure driven by ventilation towers.
Jubilee Campus, Nottingham University, 1999.
Architect: Hopkins Architects.

can be predetermined and controlled to a high degree according to need. The purpose of making a hybrid of the two systems is to reduce energy consumption whilst maintaining a comfortable and healthy indoor climate.

Hybrid ventilation combines the best attributes of the two principles (mechanical and natural) to ensure both good air quality and thermal comfort. It is a two-string system that uses the different properties of natural and mechanical systems at different times of the day or the seasons. The systems' interaction depends on both weather and usage. The hybrid ventilation system differs from a conventional system in that it features an intelligent control system, which can change automatically between natural and mechanical operation.

The many combination options offered by hybrid ventilation can lead to better indoor climate, greater user satisfaction and reduced energy consumption. The ongoing development of façade engineering that integrates ventilation systems into architectural design requires increased attention on the part of architects to the potential of new technology. As large parts of the ventilation system are now integrated into the façade, the overall concept and architectural appearance of the building alters. Instead of having lifeless façades of fixed elements and tinted glass, new dynamic architectural possibilities are emerging to articulate and animate buildings.

The need for the development of building-integrated ventilation is immense in order to tackle global warming; however, exploiting the combination options of hybrid systems places great demands on the architects and engineers involved. It is important that the architect and the engineer maintain a close dialogue during a project's creation in order to ensure creative, resource friendly solutions with conceptual cohesion between technical systems, building façades, indoor climate and architecture.

Fig. 8.35 a-b
PIHL & SON
The concept and cross section of this office building are based on natural ventilation, which can be adjusted and controlled mechanically. Night cooling of the building takes place through airing (night-time purging). The small windows placed high up are opened mechanically with apertures in a centrally located skylight. This is a typical relatively simple and well-functioning hybrid ventilation system.
Office building for Pihl & Son. Lyngby, 1994.
Architect: KHR Arkitekter.

Light and Shadow

Nanet Mathiasen
and Nina Voltelen

Light is a prerequisite for our ability to see and experience the world around us. Light describes the surroundings on the basis of the variation of light intensities that reach our eyes. Light and shadow tell us about form, materials, softness and hardness, lightness and weight. Designing buildings is to work with architectural form and light. To work with the light aperture is to design not only the room's lighting, but its appearance and mood. The interior is the reverse side of the exterior and the place where the atmosphere and character of the room is formed; it is here that the qualities of light find their expression. The light aperture is not merely a communicator of exterior light but probably the most important element in the planning of a room's visual environment.

The façade works as a light filter, which controls incident light and determines the outward view by means of the apertures. This means that the façade's aperture contributes to the creation of variations in light and shadow, so that the eye perceives details and colours, making it possible to recognise form and objects. Architectural choices are often made in relation to the local light conditions, so that the façade not only creates the room behind but also constitutes an important element in the control of the room's climate and light levels.

Over millions of years, man has developed to function optimally in relation to different light conditions on Earth. Apart from using light to see and understand the surrounding world, man also needs direct contact with light at physiological and psychological level. Our daily rhythm of rest and activity is controlled by light and our mood level is also affected. In Scandinavia, approx. 5% of the population suffers from winter depression because of the lack of natural light during the winter period. As you move further to the north, more people suffer from this ailment, whereas it is less known in the Mediterranean countries.

If we do not receive enough light in general, we may acquire illnesses such as vitamin D deficiency. Vitamin D is formed in the skin when it is irradiated with sunlight. A lack of vitamin D means that the amount of calcium produced in bones and teeth is insufficient. The bones go soft, and defects may occur in the teeths' enamel. So putting aside depression, light levels are also important to health levels. However, too much solar radiation may also be a bad thing, as too large amounts of ultraviolet light in particular can damage the skin and increase the risk of skin cancer. The higher the light intensity, the more important it is that the skin is protected – either by the pigmentation layer that forms naturally in the skin under the influence of sunlight, or by the skin being covered.

Traditionally, man has had to adapt to the specific conditions of life under the different altitudes of the Sun, both in relation to buildings and to clothing. Such adaptability has helped to create a comfortable environment around the body in an interaction between three layers – building, clothing, skin.

Fig. 9.1
SUNLIGHT IN ROOM
Museo dell' Ara Pacis, Rome, Italy, 2006.
Architect: Richard Meier & Partners.

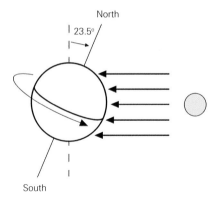

Fig. 9.2
ALTITUDE OF THE SUN
The axis of the Earth is inclined approximately 23.5°
and is hit by parallel solar rays, and as a result, the
intensity of light varies depending on where on the
globe you are. These variations have led to the
development of distinctive architectures for different
regions of the world.

Fig. 9.3
DIFFUSION OF LIGHT
The Sun's path through the atmosphere.

Light is a small part of the electromagnetic radiation that hits the Earth. It follows the laws of physics and has its own inherent characteristics. Light behaves in a predictable way, and if you know how to work with it consciously, it is possible to create functional and beautiful architecture. Light can be described by means of its three main variables: intensity, colour and direction.

As the Earth is hit by parallel rays from the Sun, the solar rays' angle in relation to the Earth's surface will differ according to the latitude, and thus the strength of the light will vary between one region and another. The solar rays that hit Earth perpendicularly have the highest intensity, and as the rays' path through the atmosphere is relatively short, the diffusion in the atmosphere is quite limited. When the Sun is low in the sky, the intensity lessens, and as the rays have to travel a long distance through the atmosphere, the diffusion of light is equally greater. When more light is spread out into the atmosphere, the sky will appear brighter. As a result, the clear sky closest to the poles is much brighter than in the areas around the Equator.

The intensity of light in a given place not only depends on the altitude of the Sun, but also on the weather conditions. The altitude of the Sun is determined by the place's location on Earth and by the season. The different seasons are caused by the approximately 23.5° inclination of the axis of the Earth in relation to the Earth's orbit around the Sun. Weather conditions determine how much of the incoming light actually reaches the Earth's surface. A light cloud cover allows a lot of light to pass through, and the sky may be so bright that it literally impairs vision, whilst a dense cloud cover can retain a large proportion of the light, thus creating a very dark sky. Heavy thunder clouds can offset the experience of time from day to night by shutting out most of the light. These dynamic conditions suggest that building façades need to be designed to accommodate climatic variation using flexible screens rather than fixed elements.

The colour of light is described as ranging from warm over neutral to cold depending on the light's spectral composition. Visible light is only a small part of the Sun's electromagnetic radiation and it is characterised according to the wavelength of the radiation. The visible light spectrum is between the 760 nm and 380 nm wavelengths (nm = nanometre = one billionth of a metre). It can be seen as colours when light is sent through a prism, which disperses the wavelengths of light differently, splitting it into narrow frequency areas.

A person with normal vision will perceive the spectrum as colours ranging from red (around 760 nm) through to orange, yellow, green and blue to violet (around 380 nm). The short (ultraviolet) and long (infrared) wavelengths are outside our visual perception range. In nature, the spectra of light are seen in the rainbow when light is broken in drops of rain. Light that contains the same amount of energy within all wavelengths in the visible area will be perceived as white. If blue colours are predominant, light will be experienced as cold, and if red colours are predominant, light will be experienced as warm. Light that contains very little energy will be perceived as black. Black and white are both abstractions and rarely occur in nature.

As it is mainly the short wavelengths (the cold colours) that are dispersed in the atmosphere, the Sun will give off a warm light, whilst the clear blue sky will give off a cool light. The light cloud cover that absorbs light form the Sun and the sky is neutral and experienced as white to grey. All three colours of light will appear on a sunny day with sunshine, blue sky and drifting clouds.

The directional intensity of the light varies from direct to diffuse. The Sun provides direct light with parallel rays, whereas diffuse light comes from all sides of the luminous firmament. In between direct sunlight and the diffuse light of the sky, there are countless variations, in which light can be more or less diffuse or more or less direct, depending on how great the dominant light source is. It may be inside, where parts of the sky are screened by the façade and its elements, or outside, when the sun is blurred by thin cloud cover or tree canopies.

The direction of light is particularly important in the shadow drawing. Direct light gives a shadow with an exact delimitation, but the greater light source in relation to that which it illuminates, the more gradual the transition between light and shadow. When the light is completely diffuse, no shadow will be seen. In architectural rendering the shadow is normally cast at 45°. Architectural drawings frequently show the shadow as black but in reality the shadow is grey and diffused by local conditions.

No matter where on Earth you are, light can be characterised on the basis of the three parameters mentioned above. They are all variable in relation to each other and dependent on physics and geometry to which the Earth, the Sun and the sky are subject. Hence, light in architecture is subtle and atmospheric as well as functional.

LIGHT AND ARCHITECTURE

Light is an important part of the internal climate and a prerequisite for many functions in the building. When the building body screens against the external climate, it also screens against daylight. Consequently, it is a balance between the external climatic conditions, the building's screening and light filtering function and the light requirement that creates the specific visual environment in a building.

If the building façade screens to an extent that daylight is insufficient for a significant part of the time, there will be an additional need to turn on artificial lighting. On the other hand, if screening is insufficient to prevent the rooms from getting too warm due to exterior solar gain, there will be a need to cool the rooms during the warm times. Both situations involve the use of electrical energy.

The building façade's possibilities of screening against and at the same time controlling sunlight and daylight are manifold. However, the optimum solution depending on latitude and thus the character of light, and hence the most appropriate design will differ from one location to another. Thorough knowledge of a place's climate and light types, coupled with careful design will ensure optimum use of light within the building for different function and location.

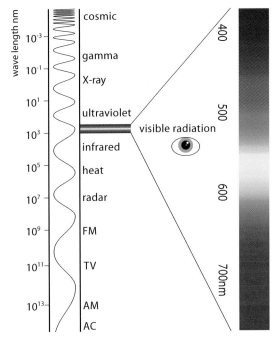

Fig. 9.4
LIGHT SPECTRUM
Electromagnetic wavelengths with indication of visible light spectrum.

The architect needs to consider three types of daylight in the planning: sunlight, skylight and reflected light.

Sunlight comes in as parallel rays directly from the Sun and is the most intensive type of light. The Sun gives a precise and hard shadow figure, which constantly changes position depending on the time of the day and the year. The high intensity of sunlight creates great contrasts, and it is often difficult to see through the shadow and out into the sunlight. Direct sunlight also brings along a lot of heat.

Skylight comes from the entire sky, which stretches from zenith down to the horizon. Skylight varies a lot in intensity, depending on whether it comes from a clear blue sky, bright white clouds or a dense dark cloud cover. In some regions, skylight is the dominant daylight source.

Reflected light comes from all the surfaces that surround us. Reflected light is always weaker than the light that hits the surface, but even so, it can be an important contribution to a building's lighting. Reflected sunlight can be stronger than the light from a blue sky. Reflected light is always present as a part of daylight.

Depending on where on Earth you are, either sunlight, skylight or reflected light can be the predominant type of light. Normally, daylight is a combination of the three but its intensity and proportion varies through the day and over the year between the types.

In dry areas around the Equator where the Sun is almost vertically above Earth all year round, the sunlight intensity is the greatest possible. The sky is relatively dark and very blue because of the limited dispersion of light through the atmosphere. Here, the Sun and the reflected light are the most dominant light source. For the sake of the indoor climate it is appropriate in this environment to keep out light and thus heat. Unscreened sunlight can heat a room to an extent where it is not comfortable to stay in it. This means that buildings are often designed with loggias, peristyles or something similar, which create shaded areas at the building perimeter. Windows are often small, and if not, it is possible to close them off completely against direct sunlight. The external reflected light is often a significant light source and hence the design of horizontal surfaces is an important consideration.

The sunlit ground reflects sufficient light to serve as the primary light source in such regions. This means that although the room is screened efficiently against the Sun, it can still be illuminated by reflected light from outside. In humid areas where the sky is covered by mist, skylight will be the most important source of daylight, even near the Equator.

Closer to the poles, where the altitude of the Sun and the weather conditions change a lot in the course of a year and even over a day, the light intensity will vary a great deal. Low sunlight spreads considerably more on its way through the atmosphere, so that the clear blue sky becomes relatively much brighter. Other climatic conditions mean that closer to the poles there is a great frequency of cloudy sky, so that for a large part of the time the sky is the most dominant light source. Here, the utilisation of skylight is very important, and the need for well lit rooms is great

Fig. 9.5
LIGHT TYPES
The three types of daylight: sunlight, skylight and reflected light.

Fig. 9.6
LIGHT IN THE BUILDING
The inclined areas around the light aperture create a gradation of light, which reduces the contrast between outside and inside. Museum of Contemporary Art, Santiago de Compostela, Spain, 1993.
Architect: Álvaro Siza.

Fig. 9.7
NORTH AFRICAN LIGHT
Close to the Equator, light is characterised by a dark blue sky and by being darkest towards zenith and brighter at the horizon.

Fig. 9.8
SCANDINAVIAN LIGHT
In Scandinavia, light is characterised by the Sun's low altitude and a bright sky.

as the climate causes people to stay indoors much of the time. Reflected light only plays a minor role when the sky is overcast. Windows will often be large in order to draw the sparse daylight into the rooms. The diffuse skylight fades into the room, giving a soft and light shadow. The characteristic description of the interior is soft transitions with a lot of slight differences in light and shadow.

As the Sun is a rare visitor, it is often welcome as a contribution to the life of the room. Due to the great variations in the altitude of the Sun, the need for screening will vary significantly from summer to winter. During summer, there may be a need to screen against both heat and glare, whilst during winter there will be a need to open up the building façade to the sparse daylight and only occasionally screen off against the low Sun or a bright, overcast sky.

LIGHT AND VISUAL COMFORT

The size and position of windows are important factors which determine where in the room light will settle, and what its character will be. Hence the detailed design of the window is of great significance to the visual conditions. The contrasts between inside and outside may be quite high. Often, the contrast can be so high that it becomes uncomfortable to look at, unless the light aperture is carefully designed.

The eye is a flexible sensory organ capable of adjusting to the great variations of light intensity on Earth. We are able to move about on bright beaches in sunshine and in dark forests at dusk and still see where we walk, but the retina of the eye needs time to adjust its sensitivity to the current amount of light. However, at a given level of light, the eye has a limited span. Too great differences in brightness cause glare, and high contrasts close to each other may result in discomfort or impaired vision.

One of the key objectives of window design is to create a gradual transition between outside and inside with variations in brightness which the eye can gradually get used to rather than be exposed to the intense light outside. Deep openings, light surfaces and areas at varying angles in relation to the light, help create visual stepping stones between outside and inside so that the glare is reduced. Dark surfaces and a lack of illuminated areas close to the window may cause unpleasantly large contrasts and hence glare. Glare often gives the impression that there is not enough light, because the eye adapts to the brighter areas.

Fig. 9.9
ADJUSTING LIGHT
A combination of external and internal shutters
supplemented by light curtains offer great
possibilities of adjusting light according to need.

Fig. 9.10
TRANSPARENT CURTAINS
In Scandinavia, it is mainly the bright sky that can cause discomfort. For generations, light curtains have been used to screen against the glare.
The Open-Air Museum, Sorgenfri, Denmark.

Fig. 9.11
FLEXIBLE SCREENING
An example of curtains in several layers that can regulate thermal comfort and also affect the light qualities in the room. Stockholm City Hall, Sweden, 1923.
Architect: Ragnar Österberg.

SCREENING AGAINST LIGHT

Buildings often consist of façades with large glass areas that let in huge amounts of light and provide a view out. However, for both climatic and visual reasons, there may be a need for additional screening and control of light on the façade's open areas. Often, it is necessary to ensure some sort of filtration of the intense sunlight at the outside, but additional screening against uncomfortable sunlight and heat may also be needed at the inside.

In places where the need for screening varies a lot, it is important that the screening is flexible. It should be possible to adapt it to various needs or have the screen removed completely. A fixed screen will always, regardless of how well designed, provide shade from some of the skylight – even during periods when it is necessary to get as much skylight into the room as possible. A fixed screen will permanently reduce the view, making flexible screening the most desirable solution. Therefore, when you want to adjust the daylight, it is essential to determine what kind of light you want to adjust. Do you primarily want to screen against sunlight and heat, or is the problem the very bright sky or reflected light from the exterior, or is there a desire to screen against people looking in and to provide a view out?

The screening function at the window can be placed outside, inside or at glass level. In general terms, screening placed outside the glass will ensure that heat stays outside and does not enter the room. Therefore, external screens are preferable in areas where the issue is specifically to keep out the heat. Interior screens, on the other hand, help keep the heat in.

If screening is placed on the inside of the glass, the heat is also brought indoors and will heat up the room. If natural ventilation cannot remove the excess heat, this may – even in temperate areas – cause difficulty in the form of overheating of the room. The main purpose of screens placed inside should be to filter light in order to create a good visual environment. Screening that is integrated into the glass should have a mainly reflective effect, as the absorption of light will cause a significant heat radiation into the room. The double façade with a naturally ventilated gap solves this problem, as heat is diverted before it reaches the room.

Light-coloured shading screening with physical depth has the advantage that light will vary across the actual screen and create a gradual transition between the dimmed light inside and the intense light outside. This makes it easier for the eye to both see inside the room and look out through the screening. Furthermore, screens with a variation in texture across the surface are much more pleasant to look at than evenly bright surfaces.

Dark surfaces on shading screens and window frames up against a bright sky will cause glare. Furthermore high contrast between the frame and sky will give an experience of darkness in the room, as the dark surfaces are what the eye sees. Screens made of perforated solid material should be fine-meshed to provide a softening effect to the great contrast between sky and screening. If the holes are too big, the high luminance of the sky will be experienced through the holes in contrast to the unlit inner surface of the screen. In this case, too, light materials will work better than dark.

Filtering

Horizontal slats can screen against the Sun by means of a given angle determined by the distance and depth of the slats. They will also screen against possible glare of the skylight, but they do not need to cover the entire window, as skylight is often only visible in the upper part. Horizontal slats are most efficient against the high solar angles around noon, whereas the low Sun is hard to screen without shutting off the window completely.

Horizontal slats make it possible to look out of the window in its entire width, and they can be designed to ensure that the immediate surroundings can be seen without significant interference. If in addition the horizontal slats are flexible, they can be a very efficient means of controlling both light, view and climate in the room.

Vertical slats primarily screen against the low Sun, whose light comes in at an angle from the side, although they are not able to reduce the intense skylight at the top of the window except by shutting off the entire window and hence also view. Vertical slats make it possible to look out of the entire height of the window, but at the same time they control the direction of both the view and the incident light depending on which way they are turned. Vertical slats should be installed so that it is possible to remove them completely from the window when they are not needed, as they will always screen out a great deal of both view and light.

A combination of horizontal and vertical slats is typical in countries where there is a need to screen against both light and heat. Such a cassette-style screen primarily allows reflected light from the surroundings through and provides a more differentiated surface to look at as a transition between outside and inside. The view is split into many small 'windows' according to the design of the 'cassettes', and this in itself may be a visual experience.

The traditional shutter as we know it from many countries consists of horizontal slats in a frame, where the frame can be opened when needed. It is a flexible, simple and efficient way to adjust light. The slats of the shutters shut out sunlight, but allow a small amount of light reflected from the ground through. It is most suitable in countries where excess light, and hence solar gain, is a problem.

Fig. 9.12
DEEP SLATS
Light deep slats provide great brightness variation as a transition between outside and inside.
Unité d'Habitation, Marseille, France, 1952.
Architect: Le Corbusier.

Fig. 9.13
SHUTTERS
In a simple way, the traditional shutter facilitates
control of both light and air.
Rome, Italy.

Fig. 9.14
DARK SLATS
Dark slats provide a great contrast to the
bright sky.
Crimp Building, Allerød, Denmark, 1998.
Architect: Vandkunsten.

Fig. 9.15
LIGHT SLATS
Light slats and glazing bars in the façade
make it easier for the eye to cope with the
transition between the light level in the
room and the skylight outside.
Sampension, Hellerup, Denmark, 2003.
Architect: 3xN.

Fig. 9.16
VERTICAL SLATS
Vertical slats control the view and the light
direction in the room to a large degree.
Arkitekternes Hus, Copenhagen, 1996.
Architect: 3xN Architects.

Fig. 9.17
POLYCARBONATE
The outer wall consists of dyed polycarbonate that obscures
the view and forms a wall of coloured light in the room.
Laban Dance Centre, London,
Architect: Herzog & de Meuron.

Fig. 9.18
COATING
In most cases, a coating of the glass that reflects heat also
adds a toning to the glass and reduces daylight penetration.
Here, sunlight is seen through an open door and one and two
façade layers, respectively, each with two layers of glass and
one layer of coating.

Diffusing surfaces

Architecturally, it may be desirable to have large, uniform glass surfaces, but this will create great problems in terms of overheating and glare in the rooms behind. Light and heat can be stopped by the glass surface to a certain extent, either by being reflected or by being absorbed. However, this will not eliminate all the unwanted heat, as the absorbed heat will be conveyed to both the room and the surroundings.

Often, a transparent coating is used that reflects the non-visible part of the solar radiation, but also a part of the visible light. In this way, a view through the entire area is achieved as well as a reduction of both incoming heat radiation and light. The disadvantage of this solution is that quite often, such a filter also adds a toning to the light that enters the room. Depending on the toning shade, this may be more or less uncomfortable, and at worst, it might affect our well-being and alter our perception of weather and the seasons.

A heavy coating may give the impression of cloudy weather, even if the sun is shining outside. In addition, this solution shuts out the same amount of light all year round, regardless of the light conditions. Normally, the entire glass area has the same filtering, which means that the high luminance of the sky in relation to the surroundings is not countered. Excess coating can lead to unnecessary energy use in the form of electric lighting.

If the window area is divided, it is possible to work with different coatings in the upper and lower fields, thus screening further against a view to the skylight. In temperate areas, a modern window will often have an internal coating, too, which is to keep heat in the room to ensure the smallest possible heat loss during winter. This coating is almost neutral.

A diffusing screen such as frosted glass and other materials that are not completely transparent allow a certain amount of light through but also spread the light, so that it is no longer possible to see through. Consequently, a movable diffusing section will not block the view permanently. As the diffusing area spreads light in all directions, it will be seen as a luminous surface. In sunlight, it can become very bright whilst at the same time the area is casting a shadow. Frosted glass and light curtains can become so bright that they dazzle more than they shield.

An area of diffusing material is often used to ensure that sunlight cannot penetrate into certain parts of the room. It may be in the form of curtains or sun-blinds, which can be drawn when the Sun shines, or it may be achieved by working with a structure on the surface, e.g. by embossing or sandblasting. By adding a silk screen printed pattern to the glass, it is possible to create a filtering of the light without necessarily changing the light's qualities. The printed pattern reflects and may transmit part of the light and allow the rest to pass through unaltered where there is no coating. This ensures that the experience of the colour and rhythm of light is maintained. Silk screen printing can be applied as a more or less transparent layer. Depending on the translucency of the paint, the printed areas can appear dark

Fig. 9.19
FROSTED GLASS
The frosted area turns dazzlingly bright when the sun shines on it, but provides shade in the room behind the façade. Stairway, The Royal Danish Academy of Fine Arts, School of Architecture, Copenhagen, 2000.
Artist: Inge Krogsgaard-Nielsen.

or highly luminous when the sunlight is direct – as is the case with frosted glass. If the fields are large and dense, there is a risk that they form an uncomfortable contrast to the skylight, whereas a fine-meshed silk screen pattern will merge and form a cohesive luminous area.

Silk screen printing can be planned to be varied across the window area and can be employed to screen against the skylight or particular areas, leaving the view unobscured. Silk screen printing can also be carried out as a coloured covering, which will then tone part of the light. Silk screen printing screens out the same amount of light all year round, and the view will also be affected at all times. However, it has many advantages and is increasingly employed as a climatic moderator on large glazed façades.

Exterior sun-blinds often consist of heavily dyed material that only allows a small part of light through. When the sun shines, they will add a coloured light to the room, which depending on the colour of the sun-blind, will also affect the experience of the room. A warm coloured sun-blind creates a golden and friendly room, whereas cold colours may contribute to the room feeling cooler and more clinical. Dark materials do not allow much light through and provide a dark surface to look at. Sun-blinds should be removable, as generally they absorb a great deal of light.

Thin, very open curtains will, to a certain extent, allow a view, as a large part of the light goes right through whilst a smaller amount of the light is dispersed from the area. Interior curtains and blinds are first and foremost a flexible solution that makes it easy to remove direct sunlight from a given area. Depending on the material's density and colour, it will change qualities of the light more or less and end up more or less bright itself. Curtains are often used in several layers, where the thin transparent curtain is used to filter the light and soften both the direct skylight and the harsh reflected light from sunlit surfaces. The thicker curtain, on the other hand, is used partly to shut out light and heat and partly to screen against cold from the window during winter or during the night. Hence, ideally both light and heavy curtains should be used.

The concept of coating can also include photovoltaic cells printed on the glass. They will retain part of the light whilst at the same time, the screened solar radiation is converted into electricity. Photovoltaic cells are available in various designs, of which the large, dense versions render a shadow pattern in the room, when the sun shines. The pattern may add sparkle or life to the room. Photovoltaic cells are also available laid out on a thin translucent membrane, where the cells are so small that they are perceived as a greyish filter. It may become possible in the future to make completely transparent photovoltaic cells that can be activated when needed. This will provide a flexible screen, where the glass area can be divided into fields that can be open or shut as needed, and the screened radiation utilised at the same time for producing electricity.

Fig. 9.21
PHOTOVOLTAIC CELLS
The canteen at The Royal Danish Academy of Fine Arts, School of Architecture, Copenhagen, is equipped with photovoltaic cells integrated into the window pane of the roof light. The photovoltaic cells have several functions: they contribute electricity to the operation, they screen out some of the sunlight, and they add an interesting pattern to the room. Architect: Fogh & Følner, 2005.

Fig. 9.20
SILK SCREEN PRINTING
Silk screen printing facilitates a great variety of individually designed decorative screens.
Town Hall, Alphen a/d Rijn, the Netherlands, 2002.
Architect: Erick van Egeraat Associated Architects.

Fig. 9.22
SHADOWS
The Sun's movement across the sky changes the shadows,
adding an extra dimension to the architecture.
Arne Jacobsen's house in Charlottenlund, Denmark, 1929.

ALTITUDE OF THE SUN AND DAYLIGHT

Man has always needed to shield himself against climate and light, and even in the greatly varied climatic situations here on Earth, different cultures have managed to adapt and create the conditions that are most appropriate in the given place. If you take a look at traditional architecture, one dominant characteristic in the planning of light apertures is consideration of the climate concerned. It is often possible to determine the latitude by looking at the design of a building's light apertures. Knowing a place's latitude, it is also possible to determine the altitude of the Sun for the different seasons. At the equinox, the Sun is right above the Equator, and the altitude of the Sun here will be 90° above the horizon at 12 noon. As the axis of the Earth is inclined 23.5° in relation to the rotation plane, at the summer solstice, the Sun will be 23.5° north of the Equator and at winter solstice 23.5° south of the Equator. The altitude of the Sun at equinox in a given place is 90° minus the latitude of the place. At the summer solstice, 23.5° are added, and at the winter solstice, 23.5° are subtracted.

Three examples

In the following, three examples are described. They are each located at specific latitudes, each have their own climate and in every way, they relate to the characteristic light conditions of the place. The altitude of the Sun in the different places is as follows:

Morocco 31°N
Equinox (90° – 31°) = 59°
Summer solstice (59° + 23.5°) = 82.5°
Winter solstice (59° – 23.5°) = 35.5°

France 43°N
Equinox (90° – 43°) = 47°
Summer solstice (47° + 23.5°) = 70.5°
Winter solstice (47° – 23.5°) = 23.5°

Denmark 56°N
Equinox (90° – 56°) = 34°
Summer solstice (34° + 23.5°) = 57.5°
Winter solstice (34° – 23.5°) = 10.5°

In the description of light levels under the individual examples, the daylight distribution is shown in lux to illustrate the visual experience and to make the examples comparable.

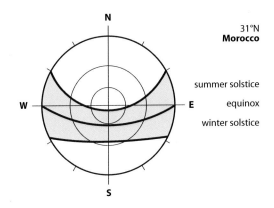

31°N
Morocco

summer solstice
equinox
winter solstice

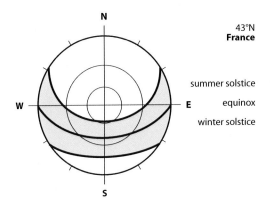

43°N
France

summer solstice
equinox
winter solstice

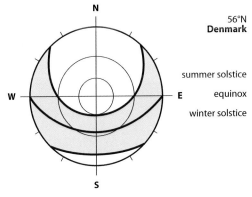

56°N
Denmark

summer solstice
equinox
winter solstice

Fig. 9.23
SUN DIAGRAMS

Fig. 9.24 a

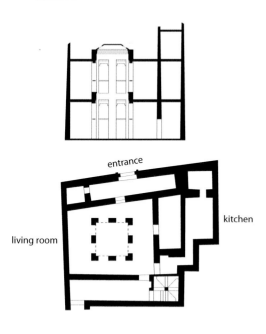

entrance

living room

kitchen

Fig. 9.24 b
SECTION AND PLAN
Plan and section through the central room, the
gallery and the small adjoined rooms. 1:500.

El Hiba, Morocco, 31°N

The village of El Hiba is situated in the southern part of Morocco. The climate here is very sunny, warm and dry. The centuries old desert village has adapted to the place's climate and is built of compact clay according to old traditions. The streets are very narrow and often built over by extra rooms from the houses. Therefore the streets become dark, cool tunnels with a view to the sky here and there. The eye is exposed to great leaps, when you move from the sunlit areas and into the alleys. However, once the eye has adapted to the low light level of the alley, the moderately lit buildings appear almost bright when you step inside them.

The houses (often with interior courtyards) are built closely together and clearly designed with the purpose of screening against the sun and creating cool rooms. The walls are up to 70 cm thick, and during the day they provide good protection against heat and the strong sunlight, and during the night, they slowly release heat to the rooms which thereby maintain a constant, pleasant and slightly cool temperature.

The traditional dwelling at El Hiba typically has two or three storeys and is built around a double-height central room with a surrounding gallery from which there is access to some smaller rooms. The central room, which serves as both living room and distribution room, is lit solely by a glazed roof light, which is covered more or less according to the season. In some of the buildings, the roof light is just a small hole through which sunlight penetrates and forms a very luminous spot on the interior walls. In others, the roof light is bigger and screened by means of straw mats and coloured crude glass, which gives a warm, soft light. The central room is the brightest in the house and it has a comfortable climate.

The gallery on the 1st floor is lit partly by the roof light and partly by the reflected light from the central room. The smaller, adjoining rooms are lit from small windows in the outer wall towards the alleys. Here, the strong reflected light from the surroundings is sent horizontally or reflected from below into the room where it hits the walls and parts of the ceiling and then softly spreads out to the rest of the room. In spite of the small windows, the rooms are well lit. At the top of the house, there is a roof terrace, where nights and evenings are often spent, weather allowing, under the cool, clear starry night.

The light level throughout the house is relatively low, but once the eye has adjusted to the low level, the rooms seem well lit for the functions they contain. A large part of the day is spent outdoors under lean-to roofs or large dark sails suspended in gardens and courtyards. Here, plenty of light is reflected from the surroundings, providing a comfortable environment.

Fig. 9.25
ROOF LIGHT
The traditional building is built around the central
room. The only light source is the partially
covered roof light.

Fig. 9.26
THE CENTRAL ROOM
The central room features a beautiful and
evenly distributed light. Although the room's
lighting comes solely from a small roof light,
the light level is sufficient.

Fig. 9.27
DRAMATIC CONTRASTS
The building offers dramatic contrasts
when you move upwards through the
dark staircase and out onto the roof.

Fig. 9.28
LIGHT LEVELS
Plan of ground floor. The plan illustrates the
light level in the central room of the house.
The light distribution is full of subtle contrasts
at a relatively low level, but as soon as the
eye has accustomed itself to the level, the
room is experienced as well lit and
comfortable.

0   6   12   25   50 100 200 400 800 1600
Approximate lux values at equinox.

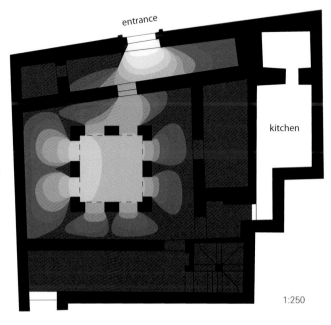

133

Fig. 9.29
BALCONIES
Facade section with balconies. The balcony serves as a permanently integrated sun screen, providing a comfortable climate for the rooms behind.

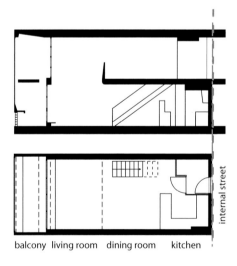

balcony   living room   dining room   kitchen

internal street

Fig. 9.30
SECTION AND PLAN
Plan and section of living room. Section of flat with living room, dining room and kitchen at the lower level. At this level, there is also access to the flat from the internal street. 1:200.

Marseille, France 43°N

Southern France has a pleasant Mediterranean climate characterised by mild winters and warm summers. Unité d'Habitation in Marseille is a large residential block designed by the architect Le Corbusier and constructed in 1952. Le Corbusier was preoccupied with the variation of the altitude of the Sun in the course of the year, and he considered this in the design of the residential block. The building contains a total of 337 flats distributed across several types of varying sizes. Typically, a dwelling consists of two floors, and via a central corridor, you enter either at the top level and walk downstairs into the rest of the flat or the other way round.

The flat shown here is designed with a kitchen, dining room and living room in linear configuration on the lower floor and bedrooms on the upper floor. The floors are connected by a staircase located in the two-storey living room. The room is lit by a large double-height window, which is slightly set back from the façade to create a balcony across the width of the room. The large window area is appropriate because the flat is deep, and the higher the window, the further the light enters into the room.

The façade is designed so as to shield the living room behind against superfluous heat. At the centre of the balcony, a shelf has been inserted, which screens out most of the sunlight when the Sun is high in the sky during summer, but allows the warmly toned reflected light to enter the room. In the winter, the low sunlight enters and heats up the room. Screening out the sunlight also implies screening out the skylight. This means that in roughly the centre of the plan there is an area with a lower light level. However, as you get away from the screening the light level rises again – precisely in the area in front of the kitchen where the dining table would typically be placed. Thanks to the screening the area right behind the façade is a bright and comfortable place to stay without direct sunlight during summertime. The lower part of the window can be opened up completely to the balcony allowing air in and a view to green areas. In this building the exterior screening helps define the interior spatialities. Not only are the rooms lit variedly, but the light contributes to stressing the functions of the different spaces.

The balcony, which serves as a permanent integrated sun screen, is a good example of Le Corbusier's *Brise-Soleil*. The system consists of three parts: a façade element that screens out direct sunlight, a glass area that facilitates a view and makes it possible to get skylight into the room, and an aperture for ventilation. It is a highly efficient system designed in relation to the Southern European climate as it screens out the high Sun that provides too much heat and at the same time allows in a large amount of the controlled skylight. The system ensures a pleasant indoor climate and provides a varied architectural expression of the façade.

Fig. 9.31
DOUBLE-HEIGHT ROOM
The double-height living room provides access to the balcony. At the centre of the balcony's double-height room, a large concrete shelf has been inserted, which is not intended for walking on, but is instead an important element in the adjustment of light conditions.

Fig. 9.32
OPEN PLAN
The living room and the kitchen are placed end to end. The room is solely lit by the large double-height window.

1:100

Fig. 9.33
LIGHT LEVELS
The lower level of a typical two-storey flat type. The drawing illustrates the room's light distribution, where an area at the centre of the room has a lower light level as a result of the façade's permanent sun screen.

0   6   12   25   50   100   200   400   800 1600
Approximate lux values at equinox.

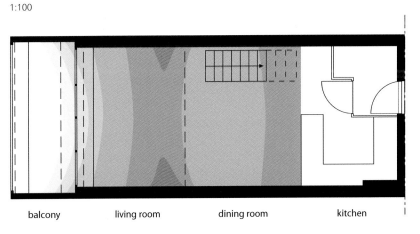

balcony   living room   dining room   kitchen

internal street

135

Fig. 9.34
PRIVATE RESIDENCE
Architect: Mogens Lassen, 1938.

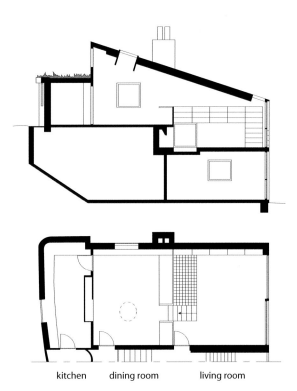

kitchen   dining room   living room

Fig. 9.35
PLAN AND SECTION
Plan and section of kitchen, dining room and living room. 1:200.

Klampenborg, Denmark, 56°N

Nordic light's coolness and weak intensity requires you to open up the interior in order to enjoy it. In 1938, architect Mogens Lassen constructed a private residence for publisher Iver Jespersen at Klampenborg in Denmark. The house is a composition of building bodies, each containing their individual function. The general characteristic is the use of large window areas that invite the skylight in. The light intakes vary from one room to another, which means that each room features its own light and mood.

To the left of the main entrance is the building body, which contains the common living areas. The dining room and the living room are placed end to end offset by half a floor. The two rooms are linked together by means of a large monopitch roof. The dining room is placed in the upper part. Here, the room is lit by a north-facing sidelight high up, an east-facing sidelight on level with the dining room, and a roof light. The high sidelight evenly distributes light from the cool northern sky into the dining room, whilst the east-facing window gathers the light around the dining table and is experienced as the room's main light source. The roof light is placed asymmetrically in the room and lifts the light level at the dining room's darkest end. Due to the shape of the roof light shaft, light is directed towards the rear wall. An extra effect has been added by painting the inside of the roof light shaft a clear shade of blue, which emphasises the skylight's colour and coolness.

In the transition from the dining room to the low-lying living room, a brick parapet separates both functions and light. In the living room, the wall contains a fireplace in front of which the sunken floor is finished with dark quarry tiles. This enhances the place as a darker area in the living room, and in particular if you move closer to the fireplace and cut off your visual field from the dining room light, the area is experienced as a special place where lighting and function underpin each other.

The living room is lit by a large glass window orientated towards the south-facing garden. The glass spans the room from wall to wall. One half features a solid panel under the window, and the other half consists of one large glass sliding door. The light intake continues approx. 1 m across the roof, which means that light is drawn further into the living room and hence the panel beneath the window is well lit.

The architect has been aware that it is appropriate to include as much sunlight and daylight as possible, so that the room appears light and airy at all times of the year. He has also achieved varied light conditions by means of the different positioning and design of the windows within the house. During the relatively short Danish summer, it is possible to screen out the sunlight by means of curtains, and the large windows facing different directions offer good possibilities for ventilation and incident solar radiation.

Fig. 9.36
WORKPLACE
The desk is placed right by the façade and directly under the roof light. Textured glass has been used for the lower part of the large sliding door and in the roof light, whilst clear glass has been used for the other glass areas. The filtering of sunlight in the room is dealt with through the use of different glass types. The textured glass diffuses the light and makes the shadows softer, whilst the clear glass allows sunlight through without obstruction, so that both shadow and the Sun's rays stand out sharply.

Fig. 9.37
SKYLIGHT
The living room with the rooflight that draws light further into the room. When standing below the roof light, you are surrounded by light and almost get the impression that you are out in the open.

Fig. 9.38
SIDELIGHT
The dining room with the east-facing sidelight where the late morning sunlight is able to enter and add heat and variation of light quality to the room.

1:100

Fig. 9.39
LIGHT LEVELS
Plan of ground floor. The plan illustrates the light distribution in the dining room and the living room. The many different windows provide an even light distribution at a relatively high light level.

0 6 12 25 50 100 200 400 800 1600
Approximate lux values at equinox.

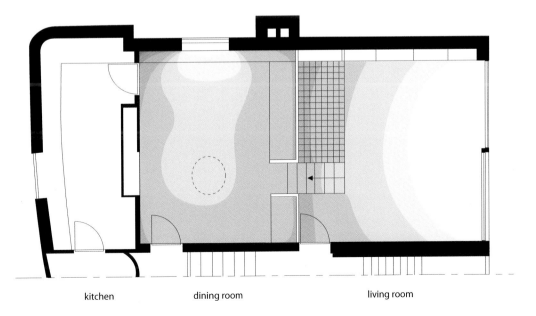

kitchen   dining room   living room

Fig. 9.40
LIGHT AS MATTER
Museum of Modern Art, New York, 2004.
Architect: Yoshio Taniguchi.

CONSIDERING DAYLIGHT

Throughout different regions of the world, there are many examples of architecture that has been adapted to the climate and the available resources. Principally, it has been a case of creating façades and climate screens where the function of the apertures is to ensure the provision of light to be able to see by and to draw in fresh air. In step with technological development, it has become possible to make the apertures increasingly larger and hence the climate screen has become more complex.

Today, there are few technical limitations to a window's size. Architecture is characterised by large glass façades, which are not necessarily large to attract light, but chosen because of a desire to achieve a particular architectural expression. The large glass areas, however, make it necessary to provide screening in order to counter the problems of too much light and heat and in winter of too much heat loss. Moreover, we turn on artificial lighting either because the rooms are experienced as too dark, or because the view to the sky spoils the eye's adjustment to the room. We cool, heat and ventilate to overcome the problems with excessive windows, without necessarily adding greater quality to the buildings.

Visual qualities emerge in the interplay between light and darkness, where shadows draw the shape and show us the position of objects. Light and shadow show us something about a materials' character, and many shades in brightness give the eyes something to rest on. Hence the qualities of light in architecture emerge, when some of the light is held back, so that other parts of the light can gain ground, thereby creating a rich interplay between space, light and surface.

The new focus on the world's resource consumption has made it necessary to design lighting in buildings so that total energy consumption is kept low. One consequence of this is that it becomes increasingly important to build buildings that allow in a sufficient amount of light for different functions. Hence the façades are designed so that daylight is distributed adequately into the room and used as the main light source for a great part of the year, and thereby reduce the demand for artificial lighting.

Light is matter to architecture just as materials, form and space are. The perception of materials, of forms and spaces is not determined solely by the material, the form, the space, but also by the light in which they are experienced. Light shapes architecture, and architecture shapes light.

Architect Mogens Voltelen,
Professor at The Royal Danish Academy of Fine Arts, 1966.

PHYSIOLOGICAL ARCHITECTURE

Physiological Architecture

Ola Wedebrunn

Climate and comfort contain both sensuous and measurable properties. It is about cold and heat, about humidity and drought, and about air movement. Climate and comfort are not fixed entities, but constantly developing both factually and as actual physical requirements and prerequisites for survival in a changing world.

The men of old were born like the wild beasts, in woods, caves, and groves, and lived on savage fare. As time went on, the thickly crowded trees in a certain place, tossed by storms and winds, and rubbing their branches against one another, caught fire, and so the inhabitants of the place were put to flight, being terrified by the furious flame. After it subsided, they drew near, and observing that they were very comfortable standing before the warm fire, they put on logs and, while thus keeping it alive, brought up other people to it, showing them by signs how much comfort they got from it.

According to the Roman architect Vitruvius (c. 80–70 BC – c. 15 BC), fire brought people together as the first climate controlling technology and at the same time it became a source for language. Vitruvius believed that architecture is both based on the knowledge of the elements earth, water, air and fire and their sensuous contrasts, and that construction of building was dependent upon climate. During the Enlightenment, however, established dogma was rejected in favour of measurable and documented science based on the close observation of nature. Even legislation and aesthetics developed related to theories about nature and climate. The Frenchman Charles-Louis de Montesquieu (1689–1755) believed that climate affects the body's physiology and that the cold parts of the world facilitate clear thinking and good work discipline and therefore presuppose different laws as compared to warm parts of the world. In contrast, the German art historian Johann Winckelmann (1717–1768) while writing *Geschichte der Kunst des Altertums*, believed that the fine arts have the best conditions in a warm climate where little clothes are needed, and where it is thus possible to study man's proportions, and that the warm climate requires less work, which gives more time to practise the arts. Like Vitruvius, the German architect Gottfried Semper (1803–1879) describes fire, the hearth, as the central comfort-creating element of the building and adds a further three elements, the terrace, the roof and the wall, as the first structure of architecture.

In recent years, the British architecture critic Reyner Banham (1922–1988) has described visions about climatic and sensuous architecture, not least in *The Architecture of the Well-Tempered Environment*, 1969. Like Semper, he bases his thoughts on the spaces formed around heat sources, but also on the functions of different climatic modifying relationships to building form.

Banham wrote the article *A Home Is Not a House*, 1965, in which the concept The 'Un-house' eliminates the traditional perception of the house, which he perceives as: *climatic adjustments and effects rather than architectural design expressions.*

Fig. 10.1
FIRE
Illustration from Marcus Vitruvius Pollio, *Tio böcker om arkitektur*, Stockholm, 1989.

Fig. 10.2
UN-HOUSE
The climate screen of the house is not wall and geometry, but merely a delimitation of service and installations. The 'Un-house' illustration, Transportable Standard-of-Living Package, The Environment Bubble, 1965. Artist: Francois Dallegret.

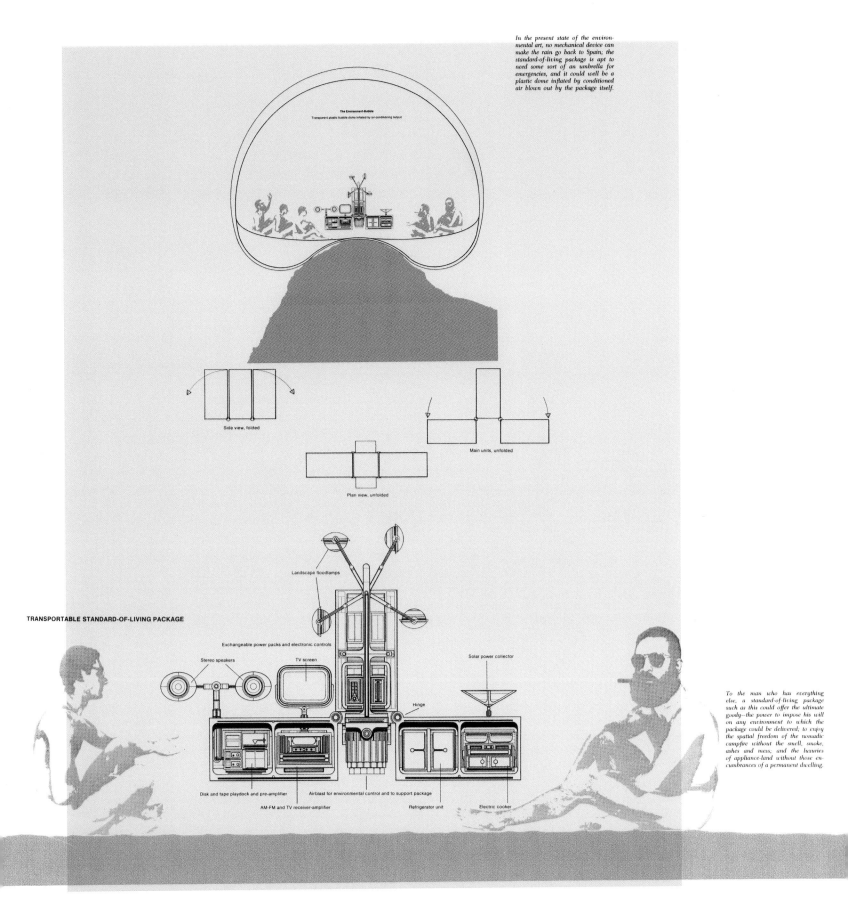

In the present state of the environmental art, no mechanical device can make the rain go back to Spain; the standard-of-living package is apt to need some sort of an umbrella for emergencies, and it could well be a plastic dome inflated by conditioned air blown out by the package itself.

The Environment-Bubble

Transparent plastic bubble dome inflated by air-conditioning output

Side view, folded

Main units, unfolded

Plan view, unfolded

Landscape floodlamps

TRANSPORTABLE STANDARD-OF-LIVING PACKAGE

Exchangeable power packs and electronic controls

Stereo speakers

TV screen

Solar power collector

Hinge

Disk and tape playdock and pre-amplifier

AM-FM and TV receiver-amplifier

Airblast for environmental control and to support package

Refrigerator unit

Electric cooker

To the man who has everything else, a standard-of-living package such as this could offer the ultimate goody—the power to impose his will on any environment to which the package could be delivered; to enjoy the spatial freedom of the nomadic campfire without the smell, smoke, ashes and mess; and the luxuries of appliance-land without those encumbrances of a permanent dwelling.

143

According to Banham, the core of the house is service and installations, whereas walls and geometry merely constitute its delimitation. Banham further refers to Buckminster Fuller's vision: *that climate will become completely controlled and the concept of a 'house' will disappear.* However, Banham did not want to completely control climate. He criticized Le Corbusier's vision of climate control as mechanically controlled buildings with a *respiration exacte* and a constant temperature of 18°C. Or as Le Corbusier wrote in 1930 in *Précisions*:

What is the basis of life? Breathing.
Breathing what? Hot, cold, dry, damp?
Breathing pure air at a constant temperature and a regular degree of humidity.
But seasons are warm or cold, dry or damp. Countries are temperate, icy, or tropical; here the 'naked man' (before London jackets) wore furs, and there he walked naked...
Every country builds its houses in response to its climate.
At this moment of general diffusion, of international scientific techniques, I propose: only one house for all countries, the house of exact breathing.

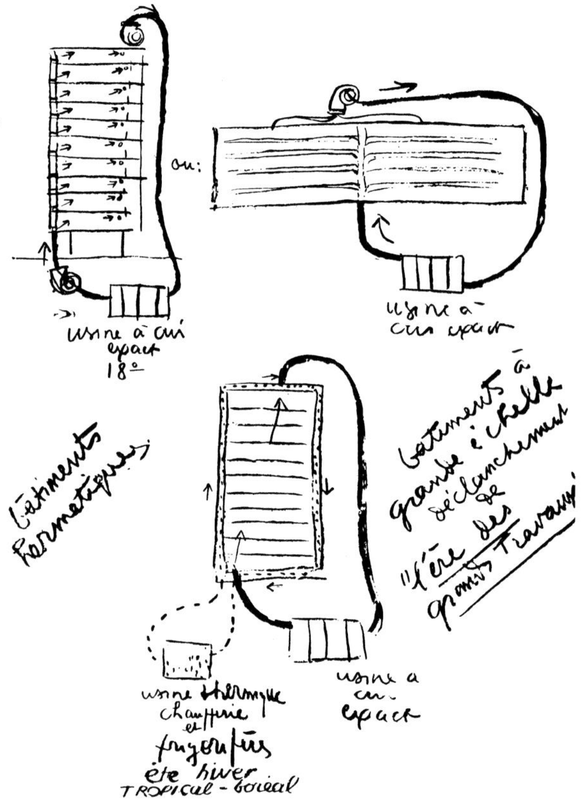

Banham criticizes Le Corbusier's idea as being determined by a desire to control and not by a desire to create well-being. Instead of the universal *respiration exacte*, Banham argued for an individually controlled climate. Just as any man can control his own radio and transport himself on his own bicycle, climate and comfort in the car can be controlled individually by means of the car's ventilation and the seat's adjustment, man should also be able to control his own building climate and his own comfort, according to Banham.

Banham describes early industrialism as *the first machine age* – where the great machines belong to industry and are the prerequisite for the products, whilst *the second machine age* produces machines that meet the needs of human beings. Machines such as the hairdryer, the food mixer, the transistor and the car have become the property of every man. Thus, the mechanically controlled standard climate of *the first machine age* is succeeded by technology of a new age directed at individual control of the indoor climate. According to Banham *the second machine age* became an age of personal control, of the machine serving mankind in the control of climate.

Although the car in many ways corresponds to Banham's ideals, architecture, design and technology are constantly developing and not tied to absolute concepts of material, scale, static principles and the laws of dynamics. Consequently, neither *respiration exacte,* Banham's 'Un-house', nor the car is the final solution to architecture and climate. In contrast to a universal vision and general ideas about climate and comfort, any room and any body constitute particular prerequisites for climatic and sensuous experience. And as such, the experience of architecture is not necessarily tied to a specific place, climate, or time.

Fig. 10.3
RESPIRATION EXACTE
Illustration from *Précisions*,
Le Corbusier, Paris, 1930.

The architect Witold Rybczynski (1943–) represents a different human science attitude to climate control and comfort. In his book *Home – A Short History of an Idea*, he presents comfort as a tradition-bound concept, into which technology is incorporated through time. According to Rybczynski, comfort in the 17th century equalled intimacy and homeliness. In the 18th century, the concept of comfort changed to mean freedom and lightness. In the 19th century, the notion of comfort was associated with mechanically supported light, heat and ventilation, and in the 20th century, it became an expression of scientific efficiency and convenience. Following Rybczynski's reasoning, one might have a guess at how the 21st century will view the comfort concept. It might give improved access to climate knowledge and handling of information. Rybczynski does not think that altered notions of comfort exclude those already in existence. He argues that we build upon pre-existing cultural and sensual knowledge.

Fig. 10.4
COMFORT
Illustration by Norman Rockwell reproduced in the book *Home – A Short History of An Idea* by Witold Rybczynski, London, 1986.

It may be enough to realize that domestic comfort involves a range of attributes – convenience, efficiency, leisure, ease, pleasure, domesticity, intimacy, and privacy – all of which contribute to the experience; common sense will do the rest. Most people – 'I may not know why I like it, but I know what I like' – recognize comfort when they experience it. This recognition involves a combination of sensations – many of them subconscious – and not only physical, but also emotional as well as intellectual, which makes comfort difficult to explain and impossible to measure.

On the one hand, Rybczynski views comfort in a historical perspective, on the other, he does not believe that it necessarily equals a development. He considers comfort both as a technological experience and as an attitude to the world and the reality of which we are a part. The experience of comfort is at any given time a complex contract, in which both historical perspective and technological and social circumstances are elements that contribute to a variation and renewal of the concept.

In *Distorsions*, the two Swiss architects Décosterd and Rahm describe in a scientific perspective how they use parameters such as hot and cold, humid and dry, light and dark to create sensuous experiences and knowledge from space and materials. Décosterd and Rahm are critical towards the general concept of climate and like neurobiologists they aim to: *assess man's nature from an organic perspective*. In connection with the exhibition *Physiologische Architektur* in Venice, 2002, they write under the title *An Endocrine Architecture*:

On the basis of this new experience, the concept of space cannot be reduced to metric and semantic dimensions. Today, it is necessary to reassess the chemical and electromagnetic elements of architectural elements in order to understand the physical connections between space and organism and to develop

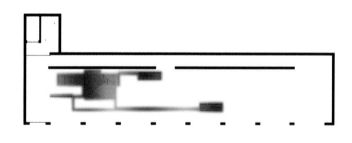

Plan

Section

Fig. 10.5 a-b
HEAT
With this illustration from 2004 for Galerie Lucy Mackintosh in
Lausanne, Décosterd and Rahm propose a space divided into zones,
which are defined by different temperatures and therefore suited
for different functions: a zone for sitting down to work at 21 °C, a
zone for visitors at 16 °C, and a zone for storage of art at 12 °C. In
the open room, pipes are laid out to provide heat during winter and
to cool during summer. Rooms are created by means of heating and
cooling from the pipes. The room's architecture is not primarily
geometric, but based upon thermal principles.

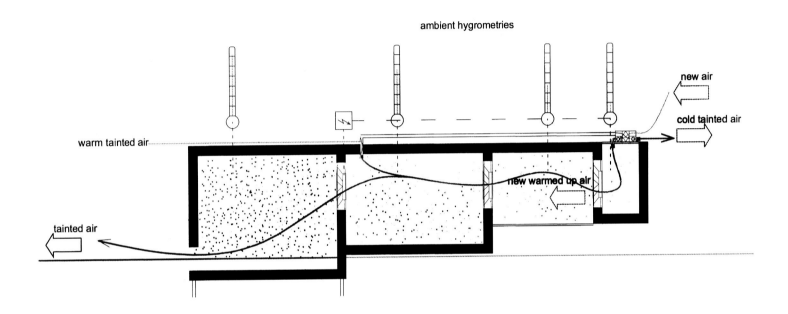

ambient hygrometries

new air

cold tainted air

warm tainted air

new warmed up air

tainted air

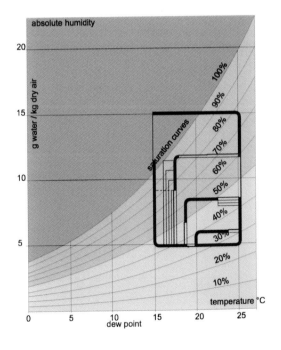

absolute humidity

Fig. 10.6 a-b

HUMIDITY

The Mollier House project by Philippe Rahm is an attempt at turning the relation between humidity and room layout into a significant architectural issue. The variation in the air humidity content from dry to humid is the starting point for the layout plan in the Mollier House.

Richard Mollier was the scientist who in the 18th century gave the world a clear and diagrammatic overview of the connection between the air's relative humidity (RH) and air temperature. Philippe Rahm has used this diagram to design a residential plan that allows air intake in the house's sauna to heat the air so that humidity drops to 30% RH. When this air seeps onwards to the bedrooms in the house, the temperature drops, and the humidity level increases to 40% RH. When the air is directed further to the bathroom and kitchen, where the temperature is even lower, the humidity level increases to 50–60% RH, and in the living room and poolroom, which are the last rooms, the humidity level increases to 70–80% RH. In this way, a sensuous and physiological cohesion is created between inhabitant and room determined by the dynamic relationship between temperature and humidity each adjusted to the need in the different rooms.

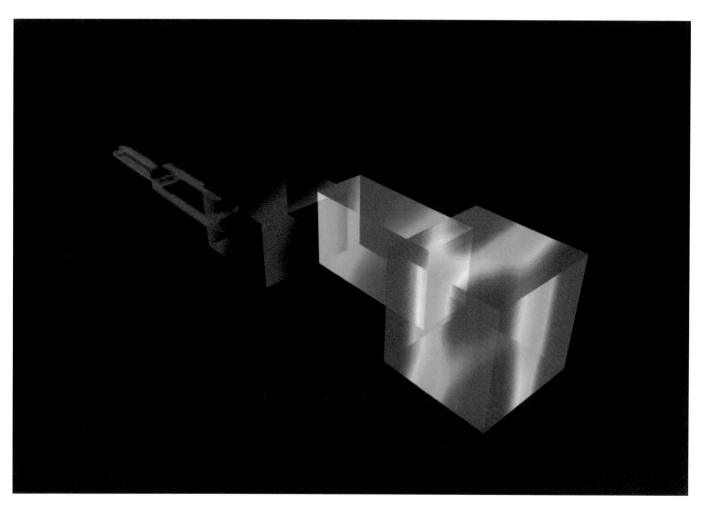

Fig. 10.7

LIGHT

With the exhibition *Nanometric spatialisation* in Paris, 2003, Décosterd and Rahm define rooms by means of light with varying wavelengths. Between 670 and 254 nm, Décosterd and Rahm construct a series of rooms that follow a linear direction, from largest to smallest, from visible to invisible, from habitable to inhabitable. A course is created, in which a space is reduced and decomposed into simpler and smaller units. Different wavelengths give different colours of light, just as they affect the skin differently. At 360 nm, UV-A waves create vitamin D and cause sunburn, and at 254 nm UV-C, germicide and ozone are formed, which destroy virus and bacterium but also attack other forms of life and are harmful to human beings.

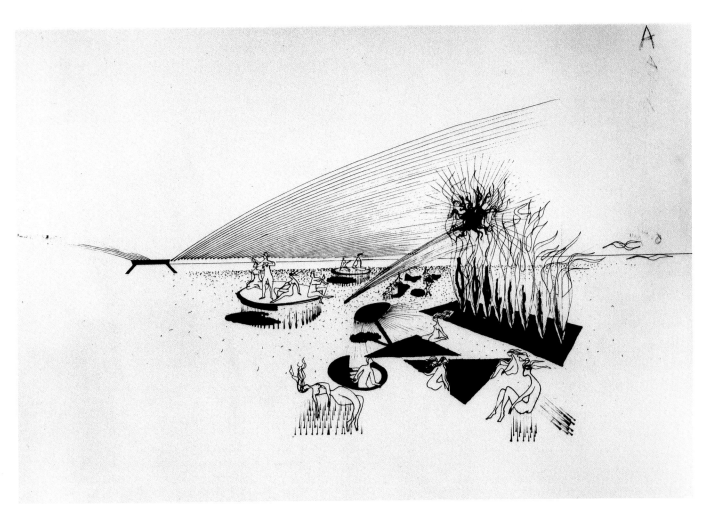

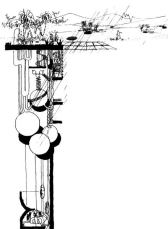

Fig. 10.8 a-b

AIR

For the past ten years I have been dreaming, as much a waking dream as possible, of a sort of return to Eden! Eden: This biblical myth is no longer a myth for me. I have always wanted to think of it in a positive, constructive, cold, and realistic way … The world of science fiction was smiling at me in its stupid, foolish way with solutions such as solar mirrors, for example, or also heating rivers in winter, creating artificial gulf streams that cross seas and oceans, changing the direction of great winds from hot countries, directing them towards cold countries and vice versa. … Of course with all the progress made by science, this is no longer an utopia today. Technique, however, could in fact realize such things! … To find nature and live once again on the surface of the whole of the earth without needing a roof or a wall. To live in nature with a great and permanent comfort.

From *Air Architecture and Air Conditioning of Space*, text and illustration Yves Klein, c. 1957.

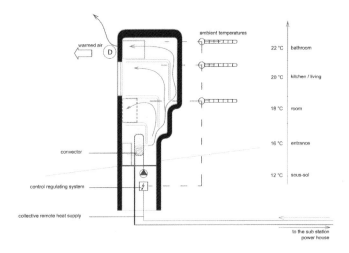

Fig. 10.9
ARCHIMEDES HOUSES
Vertical section of house with heating system
based on thermal lift. The house is divided
into areas adapted to the different functions
and needs. The air's properties, its density, its
temperature and its movement, are the
starting point for the layout of the house.
Architect: Philippe Rahm architects.

*the architectural field from the definition of the outer space to the body's organic
stimulation. Today, we should not exclude the corporeal influence from external
objects. It is becoming ever harder to delimit an artwork to the space outside the
body – solely to influence the information given outside the body and forgetting
the information's internal functional possibilities. The knowledge of the neural and
endocrine mechanisms that influence our organism offers us a significant field of
action that has never before been present in man's evolution. As we no longer
concentrate on the surface, but on the body's internal needs and on stimulating
the endocrine system, we attempt to define emotional climate data in which man
is present in a physiological sense.*

Décosterd and Rahm work with exhibitions and installations about the climate's
significance for the human body and architecture. In one of their projects, they
connect a ventilation system to a container filled with absinth, whereby the in-
jected air is turned into an intoxicating experience that affects both body and
space. In another project for a sports hall, they connect plants to the ventilation
system. The plants receive carbon dioxide from the athletes, and this is trans-
formed into oxygen, which stimulates the athletes, who in turn transform oxygen
into carbon dioxide in a balanced cycle. They also work with heat-controlled instal-
lation zones that define rooms by means of temperature, humidity, light and not as
physically tangible structures.

One of Décosterd and Rahm's sources of inspiration is the American neurologist
Antonio R. Damasio, who sees a direct connection between spirit and matter, and
believes that there is a more direct relationship between material properties and
sensuous experiences. In the same way, Décosterd and Rahm indicate an architecture of physiological and endocrine experiences. This is an architecture that takes
place outside as well as inside body, matter and space.

Décosterd and Rahm believe that physiological architecture is not tied to general
nature or artificial transformation of place and climate, but that it takes place at the
limits of the possible. According to Décosterd and Rahm, physiological architecture
is both natural and modern. It is time and space in motion, renewal, acceleration
and contraction that exceed traditionally measurable frameworks. In this sense
their work offers us a reading of architecture as construction of climate, while the
avant-garde of contemporary architecture merely explore technology as geometric
constructions.

*The current phenomena of globalization and climatic irregularity accentuate the
drift of man-oriented space into an autonomous space/time-frame, outside the
natural astronomical and meteorological rhythms. This is the perpetual spring of
the mythical Ogygia which is gradually being unfolded and elongated until they
form a kind of global climatic continuum, beyond biblical cycles, with neither sleep
nor season, night nor winter, rain nor cold. The information is instant, the*

connections simultaneous, the network is global, and uninterrupted. Here and now, but also there and tomorrow, somewhere around 21°C, at a relative humidity level of 50% with a brightness of 2000 lux, just like a fine spring day which you have decided to repeat ad infinitum, everywhere and forever and ever.
Dé costera and Rahm, *Distorsions*, Orléans, 2005.

In contrast to *respiration exacte* and *the first machine age* and the individually controlled climate in *the second machine age*, both Rybczynski and Décosterd and Rahm's offer perceptions of climate and comfort that are based on individual sensuous experience. In this sense climate and comfort are experiences and perceptions that vary in time and space from normalities to limit values and aesthetic extremes. They may – as in Vitruvius – be based on concepts of the elements of earth, water, air and fire, perceived as material properties, or on mechanically controlled properties that are measurable by means of tools as in the case of Le Corbusier. They may be meteorological and physiological as well as poetic expressions. These perceptions open up fresh possibilities for a new approach to a physiological, biological and climatically conscious architecture. One perhaps able to satisfy natural and spiritual demands of the 21st century.

Fig. 10.10
CLIMATIC MANIPULATION
With the exhibition pavilion Blur for the Landesausstellung Expo 2002 in Switzerland, the American architects Ricardo Scofidio and Elisabeth Diller created a real cloud as a climatic structure and a sensuous experience. A 140 m long bridge leads into the cloud above the water at Lac Neuchâtel.

As you enter, visual and acoustic references are slowly erased, leaving only a visual whiteout and the white noise of the pulsating nozzles. Sensory deprivation stimulates a sensory heightening: the density of air inhaled, the lowered temperature, the soft sound of water spray and the scent of the atomized lake water all begin to overwhelm the senses, inducing feelings of disorientation and isolation.

Catherine Slessor, 'Blurring Reality'
in *Architectural Review*, Sept. 2002

ILLUSTRATIONS
BIBLIOGRAPHY
INDEX

ILLUSTRATIONS

Where nothing else is indicated,
drawings: © Institute of Architectural Technology,
School of Architecture, The Royal Danish Academy of
Fine Arts.

Preface

Fig. 0.1   London City skyline.
            Photo: Jens V. Nielsen.

The Architectural Potential of Climate

Fig. 1.2   Photo: Jens V. Nielsen.
Fig. 1.3   Based on Rapoport, A., 'Vernacular de-
            sign as a model system'. In: Asquith, L.
            and Vellinga, M., *Vernacular Architecture
            in the Twenty-First Century*, Taylor &
            Francis, London/New York, 2006.
Fig. 1.5   Photo: © Bruno Klomfar.
Fig. 1.6–7 Photo: Jens V. Nielsen.

Place and Climate

Fig. 2.0   Kastrup Seaside Bathing Platform.
            White Arkitekter AB, 2004.
            Photo: Torben Dahl.
Fig. 2.1   Satellite photo: © 2009 EUMETSAT.
Fig. 2.2   Illustration from: Johannes de Sacro
            Busto, *Sphaera mundi*, Venice, 1499.
            The Royal Library, Copenhagen.
Fig. 2.5   Based on: Andersen, T. *et al.*,
            Geografihåndbogen, 2nd edition, G.E.C.
            Gads Forlag, 2000.

Human Comfort

Fig. 3.1   Photo: © Margherita Spiluttini.
Fig. 3.3 a Photo: Johannes Rauff Greisen.
Fig. 3.3 b Based on: Küppers, U. and Tributsch, H.,
            Verpacktes Leben – Verpackte Technik,
            Wiley-VCH Verlag, Weinheim, 2002.
Fig. 3.4   Photo: Torben Dahl.
Fig. 3.5   Photo: © 2003 Barry Halkin.
Fig. 3.6   Photo: © Schultz Henry Pierre.

Fig. 3.7   Drawing from: Olgyay, V., *Design With
            Climate, Bioclimatic Approach to
            Architectural Regionalism*, Princeton
            University Press, New Jersey, 1963.

Traditional Climate-adapted Architecture

Where nothing else is indicated, photo: Georg Rotne.

Fig. 4.0   Jean-Marie Tjibaou Cultural Centre
            Noumea, New Caledonia, 1998.
            Architect: Renzo Piano.
            Photo: Architekturphoto © Arcaid/Alamy.
Fig. 4.4 a-b Illustration by Hassan Fathy. Published
            in: Fathy, H., *Natural Energy and
            Vernacular Architecture*, The University
            of Chicago Press, Chicago, 1986.
Fig. 4.10 a-b Drawings from: Kawashima, C., *Japan's
            Folk Architecture, Traditional Thatched
            Farmhouses*, Kodansha International,
            Tokyo, 2000. © Kodansha International.
            Reproduced by permission. All rights
            reserved.
Fig. 4.13   Photo: Katja Bülow.
Fig. 4.14   Photo: Nanet Mathiasen.
Fig. 4.17   Photo: Katja Bülow.
Fig. 4.19 a-b Drawings from: Martínez Suárez, X.L.,
            *As galerias da marina a Coruña 1869–
            1884*, Colexio official de arquitectos de
            Galicia, Santiago de Compostela, 1987.

Large Climate Screens

Where nothing else is indicated, photo: Georg Rotne.

Fig. 5.4   © Société Belge de Photographie Ixelles-
            Bruxelles (S. III 100752 plano).
Fig. 5.7   Courtesy, The Estate of R. Buckminster
            Fuller.
Fig. 5.8   Photo: Torben Dahl.

Climate Themes

Fig. 6.0   Försterschule, Lyss, Switzerland, 1996.
            Itten & Brechbühl Architekten.
            Photo: Jens V. Nielsen.

Hot and Cold

Where nothing else is indicated, photo: Torben Dahl.

Fig. 6.1	Photo: schmidt hammer lassen architects.
Fig. 6.4	Source: Winslow, C.E.A. and Herrington, L.P., *Temperature and Human Life*, Princeton University Press, Princeton, N.J., 1949. Reproduced in: McCormick, E.J., *Human Factors Engineering*, McGraw-Hill, New York, 1964.
Fig. 6.5	Photo: NASA Headquarters – Greatest Images of NASA (NASA-HQ-GRIN).
Fig. 6.6	Photo: © The Danish Arctic Institute. Photo: J.C.A. Petersen, the Amdrup Expedition to East Greenland in 1898–1900.
Fig. 6.8	Photo: © The Hirschsprung Collection.
Fig. 6.10 a-b	Illustration: © Martín Ruiz de Azúa.
Fig. 6.13	Photo: Martin N. Johansen.
Fig. 6.14	Photo: Jussi Tiainen.
Fig. 6.16	Source: Mick Pearce, 1996.
Fig. 6.17 a-b	Photo: Thomas Yeh.
Fig. 6.18 b	Photo: Georg Rotne.
Fig. 6.19	Drawing: Ande Lundgaard, Sørensen and Rothe.
Fig. 6.21	Photo: Jan Kofod Winther, aerial photographer.
Fig. 6.23	Photo: Peter Sørensen.
Fig. 6.25 a	Photo: Jens V. Nielsen.
Fig. 6.26	Model photo: Boase.

Humidity and Precipitation

Where nothing else is indicated, photo: Peter Sørensen.

Fig. 7.1	Photo: Georg Rotne.
Fig. 7.6	Photo: Torben Dahl.
Fig. 7.7	Based on: Andersen, T. *et al.*, *Geografihåndbogen*, 2nd edition, G.E.C. Gads Forlag, 2000.
Fig. 7.9	Photo: Cardozo Yvette, Index Stop/Polfoto.
Fig. 7.10	Photo: © Big Ben. Artist: AnnaSofia Määg.
Fig. 7.11	Illustration: © Craft of Scandinavia AB.
Fig. 7.17	Photo: CINARK.
Fig. 7.18	iStockphoto.
Fig. 7.22 b	Photo: Jens V. Nielsen.
Fig. 7.24	Photo: Jens V. Nielsen.

Wind and Ventilation

Where nothing else is indicated, photo: Peter Sørensen.

Fig. 8.1	Photo: Jens V. Nielsen.
Fig. 8.3	Photo: Guy de Vuyst, DPA / Polfoto.
Fig. 8.4	Photo: Jens V. Nielsen.
Fig. 8.5	Illustration: © BDSP Partnership, www.bdsp.com.
Fig. 8.6 a	Photo: Georg Rotne.
Fig. 8.6 b	Drawing from: Alsop, W., *Le Grand Bleu – Marseille*, Academy Editions, London, 1994. © William Alsop.
Fig. 8.7	Photo: Liza Aday, iStockphoto.
Fig. 8.9 a-b	Photo: Lena McNair.
Fig. 8.9 c	Map from: Engel, J., *Huse i Sønderho* (compendium), Byggeteknisk Højskole, Copenhagen, 1974.
Fig. 8.11	Photo: Roland Jung/VisitNordjylland.dk.
Fig. 8.12	Photo: Neville Morgan/Alamy.
Fig. 8.13	Photo: Torben Dahl.
Fig. 8.14 a	Photo: Georg Rotne.
Fig. 8.18	Sketch by Glenn Murcutt. From: *Arkitekten* 28, November 1999.
Fig. 8.24-25	Photo and illustration: Dissing+Weitling architecture.
Fig. 8.26 a	Photo: Jens V. Nielsen.
Fig. 8.26 b	Photo: Torben Eskerod.
Fig. 8.27	Photo: Rob Marsh.
Fig. 8.28 a-b	Photo and illustration: Dissing+Weitling architecture.
Fig. 8.29-30	Photo and illustration: John Andersen.
Fig. 8.31	From: Rudofsky, B., *Architecture Without Architects*, Museum of Modern Art, New York,1964. Credit Atlantis Verlag, Zurich.
Fig. 8.32	Photo: Rob Marsh.
Fig. 8.33 a	© Foster + Partners. Concept sketch by Norman Foster.
Fig. 8.33 b	Photo: © Ian Lambot / Arcaid.
Fig. 8.34	Photo: Rob Marsh.

Light and Shadow
Where nothing else is indicated, photo: Nanet
Mathiasen and Nina Voltelen.
Fig. 9.16   Photo: Jens V. Nielsen.
Fig. 9.20   Photo: Torben Dahl.

Physiological Architecture
Fig. 10.0   Illustration © Philippe Rahm architects.
Fig. 10.1   Illustration from Vitruvius, M.P., *Om
               arkitektur, Tio böcker Buggförlaget*,
               Stockholm, 1989.
Fig. 10.2   Illustration by François Dallegret: Un-
               House. Transportable Standard-of-Living
               Package/The Environment Bubble. First
               published in the article 'A Home Is Not a
               House' by Reyner Banham in: *Art in
               America*, April 1965.
               © François Dallegret/billedkunst.dk.
Fig. 10.3   Illustration: Usine a air exact, Le
               Corbusier, Croquis de conférence 1929,
               Plan FLC 33527.
               © Le Corbusier/billedkunst.dk.

Fig. 10.4   Illustration: Willie Gillis in College from:
               Rybczynski, W., *Home – A Short History
               of an Idea*, London, 1988. First published
               in the *Sunday Evening Post*, October 5,
               1946. Printed by permission from The
               Norman Rockwell Family Agency.
               Copyright © 1946 Norman Rockwell
               Family Entities.
               Photo: Norman Rockwell Museum,
               Stockbridge, Mass.
Fig. 10.5 a-b Illustration © Décosterd and Rahm.
Fig. 10.6 a-b Illustration: © Philippe Rahm architects.
Fig. 10.7   Illustration: © Décosterd and Rahm.
Fig. 10.8 a-b Drawings by Yves Klein about 1957.
               First published by Daniel Herman in *Art
               Forum*, May 2004. Courtesy, Yves Klein
               Archives, Paris.
               @ Yves Klein/billedkunst.dk.
Fig. 10.9   Illustration: © Philippe Rahm architects.
Fig. 10.10   Photo: Diller Scofidio + Renfro.
Fig. 10.11   Blur Exhibition Paviliion, Expo 2002.
               Architect: Diller Scofidio + Renfro.
               Photo: Diller Scofidio + Renfro.

BIBLIOGRAPHY

General

Alberti, L.B., *On the Art of Building in Ten Books*, Rykwert, J., *et al.* (ed.), The MIT Press, Cambridge, Mass., 1988.

Behling, S. and Behling, S., *Sol Power. The Evolution of Sustainable Architecture*, Prestel Verlag, Munich/ New York, 1996.

Beim, A., Larsen, L. and Mossin, N., *Økologi og arkitektonisk kvalitet*, The Royal Danish Academy of Fine Arts, School of Architecture Publishers, Copenhagen, 2002.

Cofaith, E. *et al.*, *A Green Vitruvius. Principles and Practice of Sustainable Architectural Design*, James & James, London, 1999.

Compagno, A., *Intelligent Glass Façades*, Birkhäuser, Basel, 1995.

Dahl, T. (ed.) *et al.*, *Facaden. Teori og praksis*. Institute of Architectural Technology, The Royal Danish Academy of Fine Arts, School of Architecture Publishers, Copenhagen, 2003.

Daniels, K., *The Technology of Ecological Building. Basic Principles and Measures, Examples and Ideas*, Birkhäuser, Basel, 1997.

Deplazes, A. (ed.), *Constructing Architecture, Materials Processes Structures. A Handbook*, Birkhäuser, Basel, 2005.

Gallo, C., Sala, M. and Sayigh, A.A.M. (ed.), *Architecture: Comfort and Energy*, Elsevier Science, Oxford, 1998.

Hegger, M., Fuchs, M., Stark, T. and Zeumer, M., *Energie Atlas, Nachhaltige Architektur*, Birkhäuser, Basel, 2008.

Herzog, T. (ed.), *Solar Energy in Architecture and Urban Planning*, Prestel Verlag, Munich, 1996.

Herzog, T., Krippner, R. and Lang, W., *Façade Construction Manual*, Birkhäuser, Basel, 2004.

Koch-Nielsen, H., *Stay Cool. A Design Guide for the Built Environment in Hot Climates*, James & James, London, 2002.

Koenigsberger, O.H. *et al.*, *Manual of Tropical Housing and Building, Part 1, Climatic Design*, Longman Group, London, 1974.

Larsen, L. and Sørensen, P., *RENARCH, Sustainable Buildings, Ressourceansvarlige Huse*, The Royal Danish Academy of Fine Arts, School of Architecture, Institute of Architectural Technology, Copenhagen, 2006.

Lundgaard, B. *et al.*, Institute of Architectural Technology, *Teknik og Arkitektur – mod en bedre byggeskik år 2000*, The Royal Danish Academy of Fine Arts, School of Architecture, Copenhagen, 1995. Special edition of *Arkitekten DK*, no. 17, Nov. 1995.

Olgyay, V., *Design with Climate, Bioclimatic Approach to Architectural Regionalism*, Princeton University Press, New Jersey, 1963.

Oliver, P. (ed.), *Encyclopedia of Vernacular Architecture of the World*, Vol. 1–3, Cambridge University Press, Cambridge, 1997.

Rudofsky, B., *Architecture Without Architects*, Museum of Modern Art, New York, 1964.

Schittich, C. (ed.), *Building Skins, Concepts, Layers, Materials*, Birkhäuser, Basel, 2001.

Vitruvius Pollio, M., *The Ten Books on Architecture*, translated by Morgan, M.H., Dover Publications, New York, 1960.

Yeang, K., *Designing with Nature: The Ecological Basis for Architectural Design*, McGraw-Hill, New York, 1995.

The Architectural Potential of Climate

Rapoport, A., 'Vernacular design as a model system'. In: Asquith, L. and Vellinga, M., *Vernacular Architecture in the Twenty-First Century*, Taylor & Francis, London/ New York, 2006.

Wehinger, R., et al., *Neubau ökologisches Gemeindezentrum Ludesch* (report), May 2006, Localised on 5 May 2008 at: www.nachhaltigwirtschaften.at/nw_pdf/0651_oeko_ gemeindezentrum_ludesch.pdf

Place and Climate

Andersen, T. *et. al.*, *Geografihåndbogen*, G.E.C. Gads Forlag, Copenhagen, 2000.

Burroughs, W.J. *et. al.*, Danish edition by Christensen, L.L. and Hartby, S., *Gyldendals bog om vejret*, Copenhagen, 2002.

Theilgaard, J., *Det danske vejr*, Gyldendal, Copenhagen, 2006.

Human Comfort

Fanger, P.O., *Thermal Comfort, Analysis and Applications in Environmental Engineering*, Danish Technical Press, Copenhagen, 1970.

Heschong, L., *Thermal Delight in Architecture*, The MIT Press, Cambridge, Mass., 1979.

Küppers, U. and Tributsch, H., *Verpacktes Leben – Verpackte Technik*, Wiley-VCH, Weinheim, 2002.

Oliver, P., *Dwellings, The House across the World*, Phaidon Press, Oxford, 1987.

Traditional Climate-adapted Architecture

Cairo

Bruant, C. (ed.), *Espace Centré. Figures de l'architecture domestique dans l'Orient méditerranéen, Les cahiers de la recherche architecturale*, no. 20/21, 1987.

Fathy, H., *Architecture for the Poor: An Experiment in Rural Egypt*, University of Chicago Press, Chicago, 1979.

Fathy, H., *Natural Energy and Vernacular Architecture*, University of Chicago Press, Chicago, 1986.

Maury, B. *et al.*, *Palais et Maisons du Caire*, Vol. I & II, Centre National de la Recherche Scientifique, Paris, 1983.

Steele, J., *An Architecture for People, the Complete Works of Hassan Fathy*, Thames & Hudson, London, 1997.

Williams, C., *Islamic Monuments in Cairo, A Practical Guide*, 4th edition, The American University in Cairo Press, Cairo, 1993.

Japan

Engel, H., *The Japanese House, A Tradition for Contemporary Architecture*, Tokyo, 1964.

Gropius, W. and Tange, K., *Katsura, Tradition and Creation in Japanese Architecture*, Yale University Press, New Haven, Conn., 1960.

Kawashima, C., *Japan's Folk Architecture, Traditional Thatched Farmhouses*, Kodansha International, Tokyo, 2000.

Taut, B., *Houses and People of Japan*, Sanseido, Tokyo, 1937.

Ueda, A., *The Inner Harmony of the Japanese House*, Kodansha International, Tokyo, 1990.

Yoshida, T., *The Japanese House and Garden*, The Architectural Press, London, 1955.

La Coruña

Ángel Baldellou, M., *Arquitectura moderna en Galicia*, Sociedad Editorial Electa España, S. A., Madrid, 1995.

De Castro Arines, X., *O libro das galerías Galegas*, Edicios do Castro, La Coruña, 1975.

Martínez Suárez, X.L., *As galerias da marina a Coruña 1869-1884*, Colexio official de arquitectos de Galicia, Santiago de Compostela, 1987.

Large Climate Screens

Hix, J., *The Glass House*, Phaidon Press, London, 1996.

Loudon, J.C., *An Encyclopaedia of Gardening*, London, 1822.

Sharp, D.V. (ed.) *Glass Architecture by Paul Scheerbart and Alpine Architecture by Bruno Taut*, New York, 1972.

Vandewoude, E. *et al.*, *Les serres royales à Laeken, Bruxelles*, Donation Royale, Brussels, 1981.

Hot and Cold

Evans, M., *Housing, Climate and Comfort*, The Architectural Press, London. 1980.

Givoni B., *Climate Considerations in Building and Urban Design*, Van Nostrand Reinhold, New York, 1998.

McCormick, E.J., *Human Factors Engineering*, McGraw-Hill, New York, 1964.

Rapoport, A., *House Form and Culture*, Prentice-Hall, Englewood Cliffs, N. J., 1969.

Richardson, P., *XS Big Ideas Small Buildings*, Thames & Hudson, London, 2001.

Humidity and Precipitation

Elder, A.J. (ed.), *AJ Handbook of Building Enclosure*, The Architectural Press, London, 1974.

Hoffmann, K. and Griese, H., *Bauen mit Holz, Form, Konstruktion und Holzschutz*, Julius Hoffmann, Stuttgart, 1966.

Mostafavi, M. and Leatherbarrow, D., *On Weathering. The Life of Buildings in Time*, The MIT Press, Cambridge, Mass., 1993.

Wind and Ventilation

Cappelen J. and Jørgensen, B., *Observeret vindhastighed og retning i Danmark – med klimanormaler 1961–90*, DMI, Copenhagen, 1999.

Heiselberg, P. (ed.), *Principles of Hybrid Ventilation* (report), Hybrid Ventilation Centre, Aalborg University, 2002.

Marsh, R. and Lauring, M., *Bolig og naturlig ventilation*, Arkitektskolens Forlag, Århus, 2003.

Slessor, C., *Eco-Tech – Sustainable Architecture and High Technology*, Thames & Hudson, London, 1997.

Light and Shadow

Hopkinson, R.G., *Daylighting*, Heinemann, London, 1966.

Kaufman, J.E. (ed.), *IES Lighting Handbook, 1981, Reference Volume*, Illuminating Engineering Society of North America, 1981.

Voltelen, M., *Belysningslære*, The Lighting Laboratory, The Royal Danish Academy of Fine Arts, School of Architecture, Copenhagen, 1969.

Physiological Architecture

Banham, R., 'A Home Is Not a House' in *Art in America*, April 1965.

Banham, R., *The Architecture of the Well-Tempered Environment*, The Architectural Press, London, 1969.

Décosterd, J-G. and Rahm, P., *Physiologische Arkitektur, Architettura Fisiologica*, Birkhäuser, Basel, 2002. Published in connection with the exhibition in the Swiss pavilion at the 8th International Architecture Biennale in Venice, 2002.

Décosterd, J-G. and Rahm, P., *Distorsions*, Editions HYX, Orléans, 2005.

Diller, E. and Scofidio, R. *Blur: The Making of Nothing*, Harry A. Abrams, New York, 2002.

Klein, Y. *Air Architecture*, MAKcenter, Los Angeles, 2004.

Le Corbusier, *Precisions,* The MIT Press, Cambridge, Mass., 1991. English translation of: *Précisions*, Paris, 1930.

Rybezyuski, W. *Home – A Short History of an Idea*, Heinemann, London, 1988.

INDEX